APPLIED OPTICS FUNDAMENTALS AND DEVICE APPLICATIONS

Nano, MOEMS, and Biotechnology

Mark A. Mentzer

CRC Press
Taylor & Francis Group
Boca Raton London New York

CRC Press is an imprint of the
Taylor & Francis Group, an **informa** business

CRC Press
Taylor & Francis Group
6000 Broken Sound Parkway NW, Suite 300
Boca Raton, FL 33487-2742

First issued in paperback 2017

© 2011 by Taylor and Francis Group, LLC
CRC Press is an imprint of Taylor & Francis Group, an Informa business

No claim to original U.S. Government works

ISBN-13: 978-1-4398-2906-6 (hbk)
ISBN-13: 978-1-138-11810-2 (pbk)

Visit the Taylor & Francis Web site at
http://www.taylorandfrancis.com

and the CRC Press Web site at
http://www.crcpress.com

Contents

Foreword

After several meetings with Ashley Gasque, my mentor at CRC Press, I was asked to edit a new series on Emerging Technologies in Optical Engineering and to author the first book in that series. Having authored a text in the Marcel Dekker series on Optical Engineering in 1990 (Marcel Dekker publisher was subsequently acquired by CRC Press), and enjoying great success with that book, I heartily agreed to both requests. Ashley suggested the book address the growing convergence of multidisciplinary fields in optical engineering, in particular nanotechnology, MEMS (microelectromechanical machined structures), and optical MEMS (MOEMS), as well as biotechnology—another strong personal interest and the subject of much of my research.

The aim and scope of the CRC series are to provide timely subject matter for practicing scientists and engineers working in the cross-disciplinary fields of optics, biotechnology, applied materials and devices, integrated optics, fiber-optic sensors, electro-optics, opto-mechanics, lasers, optical computing and signal processing, image analysis, and test and evaluation, and to expand upon the growing convergences of optical engineering and related disciplines. Material is presented in a manner immediately accessible to the practicing scientist or engineer. Science, engineering, and technology continue to exhibit exponential growth in materials, applications, and cross-disciplinary relevance. This new series on Emerging Technologies in Optical Engineering serves to further amplify the growth and progress of the discipline. The first book of the series is the textbook titled *Optical Engineering: Nano, MOEMS, and Biotechnology.*

This book covers a broad range of topics encountered during my 30 years working in the field of optical engineering. Suitable as both a reference tool for the practicing scientist or engineer and as a college or university text, it provides a broad overview of the multidisciplinary fields involving optics. Timely chapters include the discussion of convergent technologies, such as MEMS, MOEMS, nanotechnology, biotechnology, fiber-optic sensing, acousto-optics (AO), electro-optics (EO), optical computing and signal processing, integrated optics, as well as specialized military applications.

Fundamental relationships, design examples, system applications, and numerous useful references are provided. This book serves as a valuable reference for industrial practitioners, researchers, academics, and students in various disciplines (including electrical, mechanical, aerospace, and materials), and for optical engineers and applied physicists and biologists interested in the interrelationships of emerging technologies and cross-disciplinary opportunities in their respective fields.

Organizations that will benefit from this book include

- Optical instrumentation and diagnostic equipment manufacturers
- Medical imaging organizations
- Biotechnology firms
- Colleges and universities
- Research laboratories and institutions
- Computer companies
- Military and the Department of Defense

This book also serves as a useful reference for venture capital technologists exploring new opportunities, investors, entrepreneurs, and those simply interested in the latest emerging technologies and applications in the marketplace. A host of application examples, fabrication and manufacturing techniques, and testing methodologies are included as additional useful material. This book is a valuable addition to the emerging fields of knowledge-driven technology in a number of scientific arenas.

Author

Dr. Mark A. Mentzer is a research scientist at the U.S. Army Research Laboratory. A native of Lancaster County, Pennsylvania, he earned his BA in physics and music at Franklin and Marshall College. He received his MSEE and PhD in electrical engineering, with emphasis on solid state physics, devices and materials, optoelectronics, and integrated optics, from the University of Delaware. He also received his MAS in business administration from Johns Hopkins University and is currently completing a MS in biotechnology, with emphasis on biochemistry and molecular biology at Johns Hopkins University.

Dr. Mentzer's current research involves laser-assisted high-brightness imaging, instrumentation for blast and blunt trauma injury model correlation, fiber-optic ballistic sensing, flash x-ray imaging, digital image correlation, image processing algorithms, applications of micro-opto-electromechanical systems (MOEMs) to nano- and biotechnology, and investigations of protein-folding dynamics. Much of his current work involves extending the fields of optical engineering and solid state physics into the realms of biochemistry and molecular biology, as well as structured research in biophotonics. This includes bioimaging, biosensors, photodynamic therapy, nanobiophotonics, biological signaling, genomics, epigenetics, and biosystems engineering.

Dr. Mentzer is the author of nearly 100 publications, 14 provisional and issued patents, 2 books, and a contributed book chapter. He serves as conference chair for numerous technical proceedings, reviews for several technical journals and publishers, frequently speaks at trade shows and conferences, and is the recipient of numerous awards for both technical and managerial excellence. He recently conducted a series of briefs to the National Academy of Sciences, National Research Council, on instrumentation and metrology for the development of personal protective equipment for the military.

During his career, Dr. Mentzer developed fiber magnetometers, integrated optic circuits, missile guidance systems, optical signal processors, magnetic memories, MOEMs telecommunication devices, laser imaging systems, biosensors, and medical imaging systems. He operated a research and aerospace defense company, managed a fiber-optic assembly manufacturing company, and ran global product development for optical systems in Europe and Asia. He also taught graduate engineering, physics, and business management courses as adjunct professor at Pennsylvania State University and Lebanon Valley College.

Dr. Mentzer lives in Maryland with his wife and two dogs. His interests include boating on the Chesapeake Bay, playing piano and organ professionally, searching out books for his library collection, and exploring the Eastern Shore.

1

Introduction to Convergent Disciplines in Optical Engineering: Nano, MOEMS, and Biotechnology

Representative examples of the evolving interrelationships of optics, nanotechnology, microelectromechanical systems (MEMS), micro-opto-electromechanical systems (MOEMS), and biotechnology include

- Digital image processing of optical imagery from high-speed cameras and x-ray illumination for real-time x-ray cineradiography
- Optical signal processing for super high-speed computing
- Physiology of the cell molecule mitochondria where messenger RNA generate proteins and signaling system pathways in response to impulse functions characterized with optical fiber sensor instrumentation
- Laser illumination with spatiotemporal filtering to see through fireballs
- Velocity and surface deformation measurement with fiber-optic Doppler sensing
- Mach–Zehnder interferometric logic gate switches for high-speed computing
- Flash x-ray cineradiographic event capture at the interface between mechanical structures and biophysiologic entities
- Nanoassembly of protein structures for protein binding to specific steps in metabolic processes

Additional applications of optical engineering to biotechnology of particular interest include bioimaging, biosensors, flow cytometry, photodynamic therapy, tissue engineering, and bionanophotonics, to name a few. Such are the applications of the future. This book is intended to extend the fields of optical engineering and solid-state physics into the realms of biochemistry and molecular biology and to ultimately redefine specialized fields such as biophotonics. It is in such crossdisciplinary endeavors that we may ultimately achieve some of the most significant breakthroughs in science.

The parallels in solid-state physics and engineering to biochemistry are enlightening. For instance, models for electron transport through respirasomes are analogous to the energetic transport of electrons in solid-state physics. The semiconductor model involves electrons in the valence band, excited to the conduction band, and essentially hopping along the crystal matrix through successive donor–acceptor energetics. In the semiconductor, "pumping" or increasing the electron energy across a band gap is necessary for electron transport, while in the biochemical model the proton-motive force across a membrane supplies the "pump."

Each mechanism exhibits its own electron transport stimuli, for instance, increased nano-ordering of water in a hydrogen-bonded contiguous macromolecule serving to "epigenetically" influence the transport of ions across a cell membrane or to influence protein folding. What are the light-activated switching mechanisms in proteins? Are there aspects of epigenetic programming that can be "switched" with light? If we "express" such a protein in a cell, can we then control aspects of the protein's behavior with light?

Consider the analogy of the progression of signal processing using combinations of digital logic gates and application-specific integrated circuits to optical signal processing using operator transforms, optical index, and nonlinear bistable functions through acousto- and electro-optic effects to perform optical computing; to Dennis Bray's "Wetware" biologic cell logic [1]; and to the astounding insights of Nick Lane in the last two chapters of his recent "Life Ascending—the Ten Great Inventions of Evolution" [2], where Lane characterizes the evolutionary development of consciousness and death (through the mitochondria), providing the basis for the biological logic and how best to modify and control that inherent logic (see Chapter 7, Section 7.5.5).

We are on the cusp of an epoch, where it is now possible to design light-activated genetic functionality, to actually regulate and program biological functionality. What if, for instance, we identify and isolate the biological "control mechanism" for the switching on and off of limb regeneration in the salamander, or, with the biophotonic emission associated with cancer cell replication, and analogous to the destruction of the cavity resonance in a laser, we reverse the biological resonance and switch or epigenetically reprogram the cellular logic?

Scaffolds are three-dimensional nanostructures fabricated using laser lithography. Perhaps, complex protein structures could be assembled onto these lattices in a preferred manner, such that introduction of the lattice frames might "seed" the correct protein configuration. This inspires the concept of biocompatible nanostructures for "biologically inspired" computing and signal processing, for example, biosensors in the military, or for recreating two-way neural processes and repairs.

By spatially recruiting metabolic enzymes, protein scaffolds help regulate synthetic metabolic pathways by balancing proton/electron flux. This represents a synthesis methodology with advantages over more standard chemical constructions. It is certainly something to look forward to in future biotechnology and biophotonics developments.

We previously designed sequences of digital optical logic gates to remove the processing bottlenecks of conventional computers. But if we now look to producing non-Von Neumann processors with biological constructs, we may truly be at the computing crossroad for the "Kurzweil singularity."

A recent article in Wired magazine entitled "Powered by Photons" [3] described the emerging field of opto-genetics in understanding how specific neuronal cell types contribute to the function of neural circuits in vivo—the idea that two-way optogenetic traffic could lead to "human–machine fusions in which the brain truly interacts with the machine rather than only giving or only accepting orders...." The fields of proteomics, genomics, epigenetic programming, bioinformatics, nanotechnology, bioterrorism, pharmacology, and related disciplines represent key technology areas. These are the sciences of the future.

References

1. Bray, D. 2009. *Wetware: A Computer in Every Living Cell.* New Haven: Yale University Press.
2. Lane, N. 2009. *Life Ascending: The Ten Great Inventions of Evolution.* New York: W.W. Norton and Co.
3. Chorost, M. November 2009. Powered by photons. *Wired.*

2

Electro-Optics

2.1 Introduction

Developing communication, instrumentation, sensors, biomedical, and data processing systems utilize a diversity of optical technology. Integrated optics is expected to complement the well-established technologies of microelectronics, optoelectronics, and fiber optics. The requirements for applying integrated optics technology to telecommunications have been explored extensively and are well documented [1]. Other applications include sensors for measuring rotation, electromagnetic fields, temperature, pressure, and many other phenomena. Areas that have received much recent attention include optical techniques for feeding and controlling GaAs monolithic microwave integrated circuits (MMIC), optical analog and digital computing systems, and optical interconnects for improving integrated system performance [2].

An example of an MMIC application [3] is a fiber-optic distribution network interconnecting monolithically integrated optical components with GaAs MMIC array elements (see Figure 2.1). The particular application described is for phased array antenna elements operating above 20 GHz. Each module requires several RF lines, bias lines, and digital lines to provide a combination of phase and gain control information, presenting an extremely complex signal distribution problem. Optical techniques transmit both analog and digital signals as well as provide small size, light weight, mechanical stability, decreased complexity (with multiplexing), and large bandwidth.

An identical RF transmission signal must be fed to all modules in parallel. The optimized system may include an external laser modulator due to limitations of direct current modulation. Much research is needed to develop the full compatibility of MMIC fabrication processes with optoelectronic components. One such device being developed is an MMIC receiver module with an integrated photodiode [4]. Use of MMIC foundry facilities for fabrication of these optical structures is discussed in Section 5.11.

Another application area that can significantly benefit from monolithic integration is optically interconnecting high-speed integrated chips, boards, and computing systems [5]. There is ongoing demand to increase the throughput of high-speed processors and computers. To meet this demand, denser higher speed integrated circuits and new computing architectures are constantly developed. Electrical interconnects and switching are identified as bottlenecks to the advancement of computer systems. Two trends brought on

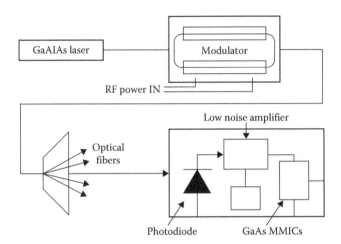

FIGURE 2.1
Optical signal distribution network for MMIC module RF signals.

by the need for faster computing systems have pushed the requirements on various levels of interconnects to the edge of what is possible with conventional electrical interconnects. The first trend is the development of higher speed and denser switching devices in silicon and GaAs. Switching speeds of logic devices are now exceeding 1 Gb/s, and high-density integration has resulted in the need for interconnect technologies to handle hundreds of output pins. The second trend is the development of new architectures for increasing parallelism, and hence throughput, of computing systems.

A representation of processor and interconnect complexity for present and proposed computing architectures [6] is shown in Figure 2.2.

The dimension along the axis is the number of processors required for various architectures. On the left side is the Von Neumann architecture

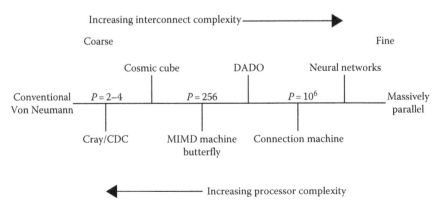

FIGURE 2.2
Processor versus interconnect complexity.

with relatively few but very complex processors. Progressing to the right, the number of processors per system increases, until it reaches a neural network requiring millions of processors, but of much lower complexity than the Von Neumann case. Looking at Figure 2.2, it becomes apparent that as the number of processors increases, the number and complexity of interconnects within the system increases dramatically. In fact, on the far right of the scale, the interconnects become an integral part of the computing architecture, and the boundary between the processors and the interconnects becomes blurred.

2.2 Optical Device Applications

The purpose of this section is to examine recent advances in GaAs technology with respect to optics applications and fabrication requirements. It will be seen that, by integrating GaAs devices and fabrication techniques into the design and development of optical circuits, many of the problems that could hinder the production of reliable and high-speed optical structures may be overcome.

Initial circuits designed were for the S to X bands, but the range of MMIC applications has now been extended as low as 50 MHz and as high as 100 GHz [7]. Frequency ranges through UHF will most likely make more use of the presently available high-speed integrated circuit silicon-based technologies than the MMIC technology. MMIC will be used in a wide variety of electronic warfare applications, such as decoys and jammers, and in phased array radars. In the commercial markets, there will be uses for MMIC technology in consumer communications' products and automotive sensors and global positioning systems. Satellite systems will be redesigned using large-scale integration and MMIC techniques to improve reliability and increase functional capacity. One of the applications of optics technology to microwave systems that has received a great deal of attention in the past few years is the use of fiber-optic modules and feed structures to replace the large and unwieldy feed structures of past phased array radar systems. The phased array technology is described in Section 2.2.1.

2.2.1 Phased Array Radar

The key item that distinguishes phased array radar from other radars is the distributed antenna configuration. The existence of an antenna array does not necessarily indicate a phased array. As defined by Liao [8], a phased array is an antenna array whose main beam maximum direction or pattern shape is controlled primarily by the relative phase of the element excitation currents. As an example, consider Figure 2.3 depicting a simple linear array

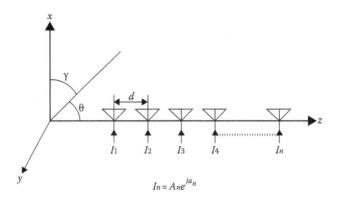

FIGURE 2.3
Excitation current relationship of a linear phased array.

arranged along the z-axis. Following the analysis in Stutzman and Thiele [9], the current in each element is represented as

$$I_0 = A_n e^{-ia_n} \tag{2.1}$$

The array is of linear phase if the elements are phased so that

$$a_n = -Bz_n \cos\theta \tag{2.2}$$

It follows from Figure 2.3 and Equations 2.1 and 2.2 that if the phase of each element is changed with time, the direction of the main beam θ is scanned. The ability to alter the direction of the main beam electronically provides the radar designer with a myriad of scanning options from which to choose.

A typical phased array radar configuration is shown in Figure 2.4, with a typical conventional radar system block diagram for comparison shown in Figure 2.5. All of the functional areas outlined in Figure 2.5 are maintained by the phased array system, but the distribution of duties has been modified. The T/R modules perform several functions that are taken care of in the beam steering, transmission, and reception functional areas in the generic radar system of Figure 2.5. The array processor performs the functions of the processor in Figure 2.5 as well as handles some of the demodulation and detection functions that were performed by the receiver. The signal generator in Figure 2.4 is accountable for the transmitter duties that are not carried out by the T/R modules. The array controller in the phased array configuration generates the phase shifting and transmits/receives switching information required by the T/R modules, and for controlling the signal generator and array processor. In addition, the array controller converts the

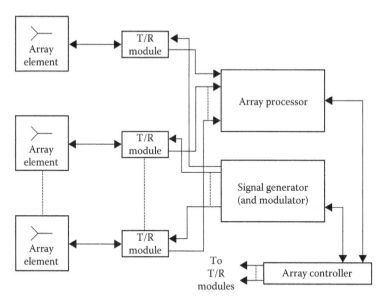

FIGURE 2.4
Typical phased array radar configuration.

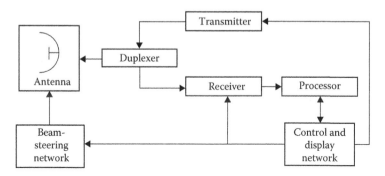

FIGURE 2.5
Block diagram of a typical radar system.

data obtained from reflected signals by the array processor into a format for display to the system operator.

The distributed nature of the radar formation in Figure 2.4 yields several operational advantages over conventional radar systems. For instance, by dividing the power amplification duties over what is typically thousands of T/R modules, solid state amplifiers may be used instead of tube-based amplifiers. This provides a decrease in size and cost per amplifier and an increase in reliability.

Another advantage realized in phased array radar designs is high-speed beam steering and radiation pattern control. Since the beam characteristics are determined by the relative current phases of the radiating elements, the array controller has complete and nearly instantaneous control over the beam direction and pattern. For the same reasons, the phased array system is an ideal candidate for computer control due to the ability to control the array beam characteristics via the phase shifters located in the T/R modules. This ability is enhanced since the phase shifters most commonly chosen for modern T/R modules are digital; therefore, the operator could run any number of phase sequencing algorithms from a computer controller to maneuver the radar beam.

The physical layout of the phased array provides some additional advantages over the nonarray radar. First, since there are typically thousands of radiating elements in a phased array radar, the system may exhibit non-catastrophic degradation. If a small percentage of the radiating elements are nonoperational, the radar will function properly but generally with an increase in peak sidelobe level [10]. Also, since phased arrays are typically built in a planar arrangement, they are more resistant to blasts, making the phased array attractive for use in hostile environments.

One of the major design decisions for a phased array radar is the choice of an array feed formation. Several options are available including the one source, one feed design implied in Figure 2.4. Figure 2.6 depicts a few other alternatives. One common implementation is the multiple-source corporate fee. An example of this is the COBRA DANE system located on Shemya Island, Alaska. This inbound and outbound missile tracking radar uses 96 traveling wave tubes to illuminate its 15,360 active elements [11]. The serpentine feed structure is being incorporated in the S-band AR320 3-D Radar [12].

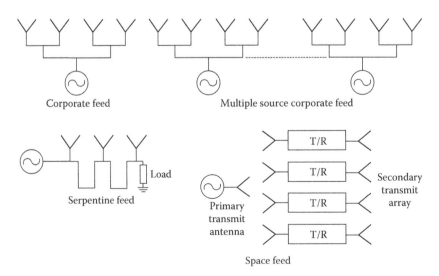

FIGURE 2.6
Array antenna feed configurations.

Another key consideration is that of operating frequency. As the radar operating frequency increases, the size of the antenna decreases, thereby enhancing the mobility of the system; however, since the antenna is smaller, either the size of the T/R module in Figure 2.4 must decrease proportionally or each source must feed a larger number of radiating elements.

The preceding considerations indicate much of the difficulty in phased radar development is mechanical in nature. One additional mechanical consideration leads to perhaps the primary reason for the lack of widespread phased array deployment. Each of the several thousand radiating elements in a phased array must be manufactured to exacting tolerances—each shifted to a high degree of accuracy. The sheer number of specialized elements makes cost the primary design driver. Tang and Brown [13] cite the following factors as the major contributors to the high cost of system development:

1. A large number of discrete components in conventional array antenna.
2. Poor production yields of high-power amplifiers.
3. High labor costs due to tight manufacturing tolerances.
4. Lack of dedicated production lines due to limited quantities.

The MMIC technology promises to overcome these problems, so that the phased array technology can come to full fruition.

2.2.2 GaAs Field Effect Transistor Technology

The advantages of GaAs are utilized in many recent developments in discrete GaAs field effect transistor (GASFET) technology. Gain and relatively low-noise characteristics were demonstrated by several manufacturers into the millimeter wave region. (In comparison, silicon devices are typically limited to applications below X-band.) In addition to its proven microwave performance, GASFET technology is readily integrated onto a common substrate with optical and optoelectronic components, permitting the development of optical monolithic microwave integrated circuit (OMMIC) systems. Since the GASFET is the basic building block of the MMIC technology, a discussion of FET structures and processes follows.

The FET is a three-terminal unipolar device in which the current through two terminals is controlled by the voltage present at the third. The term "unipolar" indicates that the FET uses only majority carriers to handle the current flow. This characteristic provides the FET with several advantages over bipolar transistors [14]:

1. It may have high-voltage gain in addition to current gain.
2. Its efficiency is higher.

3. Its operating frequency is up to the millimeter region.

4. Its noise figure is lower.

5. Its input resistance is very high, up to several megaohms.

Unipolar FETs come in two basic forms: the p-n junction gate and the Schottky barrier gate.

The first FET structure to be discussed is the p-n junction FET, or JFET. The basic physical structure of the JFET is shown in Figure 2.7a [15]. The n-type material sandwiched between two p^+-type material layers acts as the channel through which the current passes from the source to the drain of the device. The voltage applied to the gate contacts of the device determines the width of the depletion region and therefore the width of the channel. A reverse bias between the p^+-type layers causes the depletion regions to cross over into the n-type channel region. Since the channel region has fixed resistivity due to its doping profile, the resistance of the channel will vary in response to the changes in the effective cross-sectional area. With the n-type channel config-uration shown in Figure 2.7a, the electrons flow from the source to the drain. This flow would be from the drain to the source for a p-type channel sand-wiched between n^+-type gate regions. Figure 2.7b [16] depicts the restriction of the conduction channel by an increase in drain voltage. As V_D increases, so does I_D, which tends to increase the size of the depletion regions. The reverse bias, being larger toward the drain than toward the source, generates the tilt in the depletion region toward the drain. Since the resistance of the restricted conduction channel increases, the I–V characteristic of the chan-nel diverges from linear. As V_D increases more, there is a point at which the value of I_D levels off and the channel is completely pinched off by the deple-tion regions. Once the device is in this saturated operating region, the drain current may be modulated by varying the gate voltage.

The JFET was the first variety of field effect device developed and is still found in many applications. Its structure, however, requires a multiple-diffusion process and so makes it a difficult device to integrate. The device is also limited to frequencies below X-band due to the slow transit times under the diffused gate regions.

The Schottky barrier gate FET is of most interest for the development of OMMICs. Schottky suggested in 1938 that a potential barrier could arise from stable space charges in the semiconductor without the introduction of a chemical layer. This provided the foundation for the development of the metal semiconductor FET (MESFET).

The physical structure of the MESFET is shown in Figure 2.8. This particu-lar GASFET is a low-noise device that has been fabricated using ion implan-tation techniques. This technique will be compared with others for the mass production of MMIC devices in Sections 5.10 and 5.11. The device fabrication process begins with the implantation of donor ions (in this case, Si^+ ions) directly into a semi-insulating GaAs substrate wafer. Next, n^+ contact layers

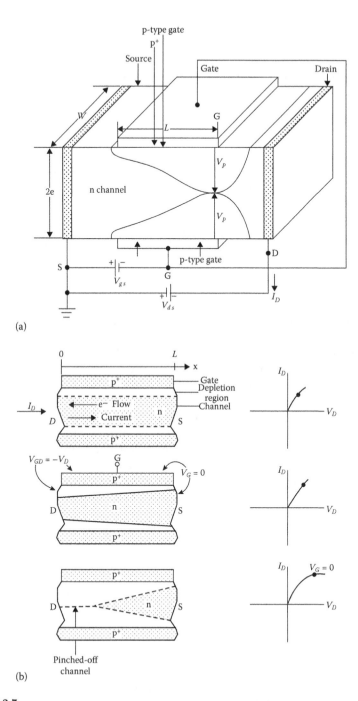

FIGURE 2.7
JFET structure and characteristics (a) p-n junction gate FET (JFET) and (b) JFET channel depletion characteristic.

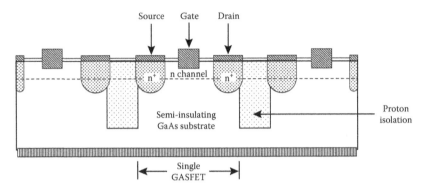

FIGURE 2.8
Integrated GASFET structure cross section. (Adapted from K. Wang and S. Wang, "State-of-the-art ion-implanted low-noise GaAs MESEFETs and high-performance monolithic amplifiers.")

are implanted into the source and drain contact regions to help reduce the source resistance. The FET source and drain ohmic contacts are formed next using gold alloys. High temperatures are used to bond the contacts and to ensure smooth edges. The contact photolithography process continues with the formation of the gate metallization. This is the most critical step of the process since the gate metallization is typically 0.5 μm long by 300 μm wide and forms the space charge, or depletion, region. The gate metallization is generally an aluminum alloy.

The MESFET in Figure 2.8 has a gate made of titanium and aluminum and then overlayered with a titanium and gold mixture for a low-resistance bonding. In MMIC fabrication, device isolation is an important factor in reducing RF losses. Good isolation and a reduction in pad capacitance are achieved, as shown in Figure 2.8, by direct proton bombardment [17].

The basic operation of the MESFET has been described by Liao [18]: "A voltage is applied in the direction to reverse bias the n⁺-n junction between the source and gate, while the source and drain electrodes are forward biased. Under this bias condition the majority carrier electrons flow in the n-type layer from the source electrode, through the channel beneath the gate, to the drain electrode. The current in the channel causes a voltage drop along its length so that the Schottky barrier gate electrode becomes progressively more reverse biased toward the drain electrode. As a result, a charge depletion region is set up in the channel and gradually pinches off the channel against the semi-insulating substrate toward the gate end. As the reverse bias between the source and the gate region increases, so does the height of the charge depletion region. The decrease of the channel height in the nonpinched-off region will increase the channel resistance. Consequently, the drain current I_D will be modulated by the gate voltage."

In the microwave domain, the MESFET design has the advantage of a very short gate length that in conjunction with the high electron mobility of GaAs results in short transit times beneath the gate region, thereby increasing the available frequency of operation for the device. Standard low-noise discrete GASFETs have gate lengths as short as 0.25 μm. Gate lengths as short as 0.1 μm have been reported in the literature, but MMIC devices are generally limited to 0.5 μm gate lengths due to process limitations and low yields.

The simple structure of the GASFET along with its superior frequency response compared to JFETs and bipolar transistors have made it the fundamental building block of MMIC technology. Heterostructure and superlattice devices may eventually replace the MESFET, but not before processing technology matures considerably. In addition to the ion implantation fabrication technology, molecular beam epitaxy (MBE) and metal-organic chemical vapor deposition (MOCVD) are used in many MMIC processes.

2.2.3 Optical Control of Microwave Devices

The control of microwave devices and circuits using optical rather than electronic signals has gained much attention in the past few years [19]. Experimental studies have been ongoing since the 1960s, but only in the past several years has the experimentation come to fruition in devices and systems. New high-speed electro-optic devices and fiber-optic distribution systems are the main reason for bringing the results of these earlier experimental studies to the applications arena, thus increasing the interest in controlling microwave devices by optical means.

The reasons for optically controlling microwave devices and circuits are many. First, microwave devices and systems are becoming more and more complicated and sophisticated. They require faster control and higher modulation rates. Using optical illumination as a source of control is one way to fulfill these requirements [20]. Optical control is faster because an optical signal does not encounter the inherent delays that an electrical signal encounters such as rise-time RC time constants in control circuitry and cabling. Second, optical control yields greater isolation of the control signal from the microwave signal. It is also much simpler to design the optical control circuitry than the electronic control circuitry because the electronic circuitry couples unwanted signals into the output if great care is not taken in the design of the optical control circuitry itself. Greater isolation improves the spurious response of the output signal as well as decreases the design complexity of the control circuitry. Both of these factors result in optically controlled devices having a better output noise specification at a lower cost. Another benefit of using optical control is that electro-optic and microwave devices can be fabricated on a monolithic integrated circuit because of the similarity in material and fabrication techniques. This results in smaller lighter weight components at a potentially lower cost once the fabrication processes are developed. Compatibility with optical fibers is another advantage of optical control. Optical fibers are becoming less costly, and are smaller and lighter than their electronic counterparts—coaxial cable and waveguide. This weight and space reduction is very important when considering transportable (man or vehicle) or airborne equipment. Finally, optical control is inherently immune to electromagnetic interference (EMI). This feature is becoming more important each day with secure transmissions and electronic warfare/countermeasures.

There are basically two types of optical control of microwave devices and systems. One is control of passive components such as microstrip lines or dielectric resonators. Examples of theses are switches, phase-shifters, attenuators, and dielectric resonator oscillators (DROs). The second type is the control of active devices such as IMPATT diodes [21], TRAPATT diodes [22], MESFETs [23], and transistor oscillators [24]. Regardless of whether one is controlling passive or active devices, optical control is governed by the following: (1) illumination of photosensitive material, (2) absorption of the illumination, and (3) generation of free carriers due to the illumination.

2.2.3.1 Optical Control of Active Devices: IMPATT Oscillators

One type of active device that has been used quite extensively in experiments to illustrate optical control capability is the IMPATT oscillator. IMPATT stands for *IMP*act *I*onization *A*valanche *T*ransit *T*ime. The IMPATT diode uses impact ionization to generate free carriers that then travel down the drift region. The avalanche delay and the transit time delay allow the voltage and current to be 180° out of phase that creates the negative resistance needed for oscillation.

Figures 2.9 through 2.11 show the basic characteristics of a Read diode. The Read diode is one of several diode types that can be used as an IMPATT diode. Other types are PIN diodes, one-sided abrupt p-n junction, and modified Read diodes. The following discussion deals only with the Read diode structure and the theory of operation.

Figure 2.9 shows the basic device structure, doping profile, electric field distribution, and avalanche breakdown region for a p^+-n-i-n^+ Read diode. Note the electric field distribution. At the p^+-n junction, the electric field is a maximum and decreases linearly until the intrinsic region is reached. The electric field is then constant throughout the intrinsic region. At the point of maximum electric field, the generation of electron–hole pairs occurs through avalanche breakdown. This region is called the avalanche region and is shown in Figure 2.9c. The free carriers travel across the region of constant electric field, called the drift region.

Figure 2.10 shows the Read diode connected across a reverse biased DC voltage, as shown in Figure 2.10d. The DC voltage is equal to the reverse breakdown voltage of the diode, so that breakdown occurs during the positive half cycle of the AC voltage avalanche; during the negative half cycle of the AC voltage, avalanche breakdown has ceased and the carriers drift at their saturation velocity. Looking at Figure 2.10e, note how the injected

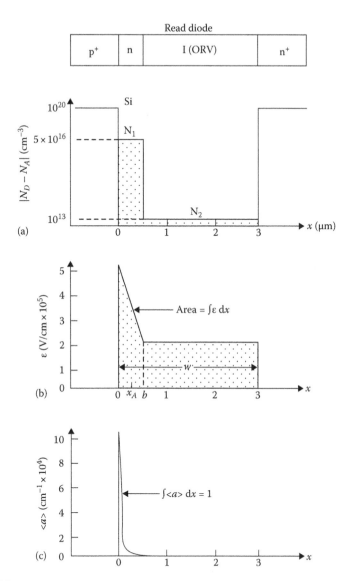

FIGURE 2.9
Read diode (a) doping profile (p-n-i-n), (b) ionized integrand at avalanche breakdown and (c) Avalanche region. (After Bhasin, K.P. et al., Monolithic optical integrated control circuitry for GaAs MMIC-based phased arrays, *Proc. SPIE* 578, September, 1985.).

current peak occurs not at the AC voltage peak ($\Phi = \pi/2$) but at the point when the AC voltage becomes negative ($\Phi = \pi$).

This is because of the avalanche phenomena. Avalanche occurs when electron–hole pairs are created. These free carriers have enough energy that when they impact with other electrons, ionization occurs, creating even

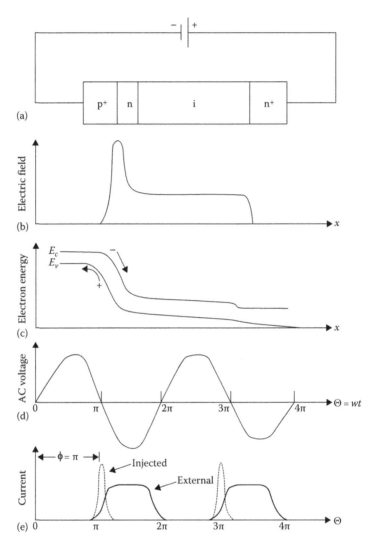

FIGURE 2.10
Read diode (a) p-n-#-n structure, (b) field at avalanche breakdown, (c) energy-band diagram, (d) AC voltage, and (e) injected and external currents (After Bhasin, K.P. et al., Monolithic optical integrated control circuitry for GaAs MMIC-based phased arrays, *Proc. SPIE* 578, September, 1985.).

more free carriers. As long as the bias voltage across the diode is larger than the breakdown voltage, avalanche breakdown occurs; therefore, the maximum amount of avalanche current occurs at the time just before the bias dips below the breakdown voltage. This is referred to as the injection phase delay.

The injection phase delay must be present for the IMPATT diode to exhibit negative resistance. This can be seen by looking at Figure 2.11 which is a plot of AC resistance at three different injection phase delays. Note that when there

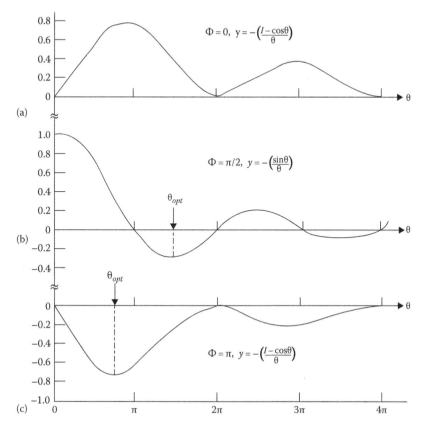

FIGURE 2.11
AC resistance versus transit angle for three different injection phase delays, (a) $\Phi=0$, (b) $\Phi=n_2$, (c) $\Phi=n$.

is no injection phase delay ($\Phi=\pi$), the AC resistance never becomes negative. Therefore, no oscillation can occur when there is no injection phase delay. The maximum negative AC resistance then occurs when the AC voltage just starts into its negative half cycle. The injected carriers drift across the drift region of length W at their saturation velocity, v_s, during the negative half cycle of the AC voltage. Therefore, the frequency of oscillation is $f=v_s/2W$.

Several important characteristics of the microwave packages used for IMPATT oscillators are the cavity and the heat sink material at the base of the package. The cavity envelops the IMPATT diode, and its dimensions are such that oscillations at the correct frequency are obtained. The packaging material must have a relatively low coefficient of expansion relative to temperature in order to ensure proper oscillation over the specified temperature range. The heat sink material must be a good thermal conductor because of the heat buildup from the high electric fields present in the avalanche region that cause impact ionization current.

The IMPATT diode is the most powerful solid state device in the microwave and millimeter wave regions. Pulsed outputs of >30 W have been achieved at 10 GHz. Output power approaching 10 W has been seen at a frequency of >100 GHz. The efficiency of the IMPATT diode has reached as high as 37% in a pulsed operational mode. The aforementioned IMPATT diode characteristics make it an important device for the future and a logical choice for optical control.

2.2.3.2 Illumination Effect on IMPATT Diode Operation

Many effects on IMPATT diode behavior due to optical control have been investigated [26]. This section will cover four of the most important effects on performance. These are frequency tuning or shifting, noise reduction, AM/FM modulation of the oscillating frequency, and injection locking of the IMPATT, which is a "cleaner" signal.

Frequency tuning is a very important capability for oscillators. Not only can the frequency be shifted by illuminating the IMPATT diode, but the shifting can be accomplished much more quickly than by conventional control techniques. Noise reduction is important in any frequency-generating system because the lower the noise, the more sensitive a system will be to small offsets of frequency such as in a Doppler radar or a coherent receiver system. Also, noise reduction is important since present frequency spectrums are already overloaded. By reducing the noise output of an oscillator, the interference with other systems operating on nearby frequencies will be reduced. AM/FM modulation of the IMPATT is very useful when using the IMPATT as an oscillator in a communications system, because now the diode can be modulated by a light source in order to transmit information from one point to another. Also, because the light source is usually a laser, the modulation frequency can be expected to go high in the future due to the constant improvements in laser technology. Finally, injection locking of the IMPATT diode using optical techniques is very promising because of the capability to control the frequency of one or more frequency sources. This is very beneficial when utilizing IMPATTs in a system where coherent sources are necessary as in Doppler radars.

2.2.3.3 Experimental Results on IMPATT Diodes: Optical Tuning

Optical tuning is achieved by illuminating the IMPATT diode by a laser source. Photogeneration of carriers occurs that alters the timing of the avalanche cycle. This alteration in avalanche injection phase produces a change in the oscillation frequency of the diode. These changes in frequency can occur more quickly than by conventional means of control.

Data from one particular experiment will now be discussed [27]. This experiment is singled out because the IMPATT diode had an operating

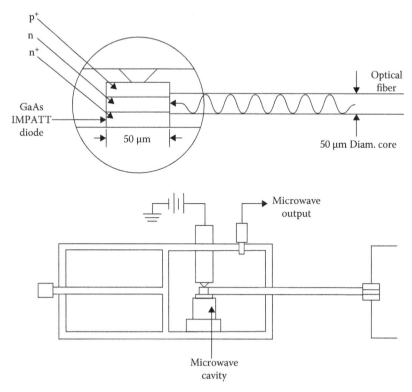

FIGURE 2.12
Optically synchronized microwave oscillator.

frequency of about 92 GHz. Most previous experimentation on optical control had been done on IMPATTs that operated below 18 GHz.

In this experiment, the p^+-n-n^+ diode was illuminated by a GaAs/GaAlAs laser operating at 850 nm. Figure 2.12 shows the basic method used to illuminate the IMPATT diode. The output frequency was first measured with the free-running oscillator. The frequency was 91.83 GHz with a bias current of 100 mA. The diode was then illuminated with 3 mW of laser power that optically generated 20.5 µA of current. This resulted in a 9 MHz shift in frequency. Less than 10 MHz frequency shift was achieved because of the poor coupling between the laser and the IMPATT diode. If the coupling efficiency was improved to the point that typical high-speed photodiodes have performed, a frequency shift of 600 MHz could be obtained.

In addition to the tuning range, tuning speed is another important parameter. Ninety percent of the 9.4 MHz frequency change occurred in 55 ps, or just <20,000 changes in frequency in about 1 µs. Optical control certainly allows fast response.

2.2.3.4 Noise Reduction by Optical Means

One source of noise in an IMPATT device is the noise generated during the avalanche multiplication process [28]. This noise is due to the variation in successive cycles of the injected avalanche current. This variation is due to the small amount of carriers present during the start-up of the avalanche cycle that leads to jitter in the avalanche cycle. This causes high oscillator noise levels. By increasing the reverse saturation current in the IMPATT, noise can be reduced. Optical illumination can be used to increase the reverse saturation current by photogenerating carriers in the diode.

Experiments were performed on both Si and GaAs IMPATT diodes. The Si oscillator had an RF output power of 260 mW at 10.47 GHz, and the GaAs oscillator power was 420 mW at 10.44 GHz. Both diodes were illuminated with a HeNe laser at 632.8 nm. The FM noise output from 0 to 50 kHz offset was measured for both diodes.

At very low illumination levels, no improvement in noise performance was measured. This is because the optically produced current is much less than the thermally generated saturation current that is greater for Si than for GaAs since the bandgap for Si is less than that for GaAs.

Increasing the illumination resulted in a decreasing FM noise level. The GaAs diode showed a 5 dB improvement in carrier/noise level at <7.5 kHz offset from carrier. The Si diode showed only a 2 dB improvement at <7.5 kHz offset because of the greater reverse saturation current already present in the Si diode. Results from this experiment also showed that if laser sources with greater amplitude stability were available, the noise performance of the IMPATT diode could be improved even further.

2.2.3.5 Optically Induced AM/FM Modulation

Optically illumination can be used to modulate an IMPATT oscillator [29]. AM modulation can be attained by two methods—quenching or enhancing. Quenching occurs when illumination of the diode alters the Q factor of the oscillator circuit enough to stop all oscillations of the diode. Enhancement of the oscillator output can occur with illumination of the diode, but only within a limited optical power range; therefore, quenching is the dominant effect used to AM modulate an IMPATT diode oscillator. FM modulation is produced by the change in susceptance of the device due to optical illumination. This change in susceptance leads to significant shifts in frequency.

Experimental results on X-band silicon IMPATT diodes that were modulated by optical means will now be discussed. The laser used was a 10 W 900 nm GaAlAs source. The laser was pulsed for 0.1 μs at a repetition rate of 200 Hz. AM modulation was observed in both the quenched and the enhancement modes. In the quenched mode, the ratio of RF power during laser-on and laser-off states was about 10 dB. A total of 20 dB was obtained in the enhancement mode; however, AM modulation in the enhancement mode is much more difficult to implement. FM modulation was also observed. Frequency deviations of >4% were seen. Greater deviation could be reached at the expense of excessive AM modulation.

2.2.3.6 Optical Injection Locking

Optical injection locking is the technique whereby the IMPATT frequency is controlled by a purer microwave source [30]. This technique is useful when many IMPATTs must operate in a coherent fashion. Optical injection locking has a very promising future in the fields of radar and communications.

There are two types of optical injection locking. One is direct injection locking and the other is indirect injection locking. Direct injection is illustrated in Figure 2.13. A laser is used to illuminate the IMPATT diode that is encased in a microwave cavity. Along the way, the laser light is intensity modulated by a microwave source feeding an electro-optic modulator. This intensity modulation triggers the avalanche process, thereby controlling the frequency of oscillation of the IMPATT. This is referred to as the direct injection locking method because the optical signal is directly illuminating the IMPATT. Indirect injection locking is shown in Figure 2.14. Note the difference between direct and indirect locking. Indirect locking the optical signal, which is modulated by the master oscillator, does not reach the IMPATT diode at all. The modulated optical signal controls an intermediate device that then electrically controls the IMPATT. In Figure 2.14, the intermediate device is the PIN diode.

Another difference between methods of injection locking is whether the fundamental frequency of the IMPATT is used as the cleaner microwave source frequency or whether a subharmonic of the IMPATT frequency is used as the microwave source frequency. For example, in Figure 2.13 [31],

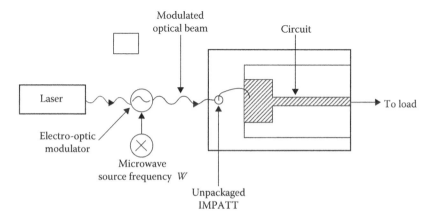

FIGURE 2.13
Optical synchronized microwave oscillator.

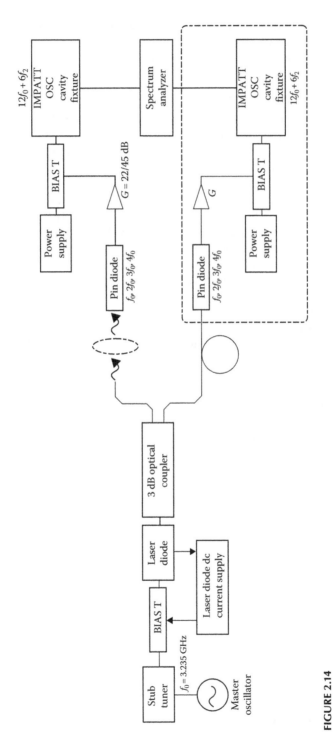

FIGURE 2.14
Experimental setup for indirect optical injection locking of 2 mm wave IMPATT oscillators. The dotted enclosed section is not presently used.

if the IMPATT frequency is 10 GHz and the microwave frequency into the electro-optic modulator is 10 GHz, then fundamental frequency locking is occurring. However, if the IMPATT frequency is 10 GHz and the microwave source frequency is some subharmonic of the IMPATT, say 5 GHz, then subharmonic locking is occurring. Subharmonic locking is advantageous over fundamental locking, especially in higher IMPATT frequency applications, because lower frequency microwave sources can be used that results in cost savings.

Experimental results of a 38.82 GHz silicon IMPATT oscillator using indirect subharmonic injection locking will now be discussed. Figure 2.14 shows the experimental setup. The master oscillator frequency is 3.235 GHz. This frequency modulates the drive current to the laser that results in the laser, being intensity modulated at the same frequency. The output power of the laser is 10 mW and it operates at a wavelength of 830 nm. The modulated laser output is split and the two laser signals control the PIN photodetectors. The fourth harmonic of the master oscillator, 12.94 GHz, is generated by the PIN photodetectors and is amplified by variable gain amplifiers. This fourth harmonic is electrically injected into the IMPATT and injection locking at 38.82 GHz, the third harmonic of 12.94 GHz, occurs.

Results of the experiment are encouraging. FM noise performance improvement of the injection locked IMPATT versus the free-running IMPATT is significant. The free-running IMPATT FM noise was −50 dBc at 100 kHz offset from carrier. Once the IMPATTs were injection locked, the FM noise was −55 dBc at 5 kHz offset from carrier. Locking range of the IMPATT was also evaluated. When the amplifier following the PIN photodetector had a gain of 22 dB, a locking range of 2 MHz was observed. With an amplifier gain of 45 dB, the locking range was 132 MHz. These results indicate that indirect injection locking could be used to apply FM to the master oscillator in order to use the IMPATT in a communication or radar system.

2.2.4 TRAPATT Oscillators

TRAPATT stands for *TRApped Plasma Avalanche Triggered Transit*. Oscillators using TRAPATT diodes have higher power and high-efficiency capability [32]. Pulsed outputs of up to 1.2 kW at 1.1 GHz and efficiencies of 75% at 0.6 GHz have been observed. Operating frequencies are limited to the microwave frequencies and below.

Diode structures can be either n^+-p-p^+ or p^+-n-n^+. The basic theory of operation is as follows. Upon application of a current step function to the diode, an electric field will be established that decreases linearly with distance from the injected current. The electric field will increase until the critical field, E_m, is reached. This critical field will move across the diode causing avalanche breakdown. The velocity with which the avalanche zone sweeps across the diode is very high-higher than the saturation velocity of the newly created free carriers. This leaves the diode filled with a plasma of free carriers. This plasma now travels across the diode, but at very low velocity due to the voltage across the diode dropping significantly after the avalanche zone passes. Because of the low velocity of the free carriers, the transit time across the diode can be much longer than if they were traveling at the saturation velocity. This transit time delay provides the needed phase delay for oscillation. Because the transit time is relatively long, the frequency of operation can be quite low compared with the IMPATT diode.

2.2.4.1 Illumination Effect on TRAPATT Operation

Three effects of illumination on TRAPATT diodes will now be discussed. These are not the only effects observed, but are effects that have real system applications. The effects are reduction of startup jitter, frequency shifting, and variation of RF power output [33].

2.2.4.2 Experimental Results: Start-Up Jitter Reduction

Start-up jitter in TRAPATT oscillators can be very troublesome. Jitter on the order of 100 ns is common and this becomes even worse with temperature. Optical injection of carriers into the diode during the low-current portion of the RF cycle can stabilize the TRAPATT mode and thus reduce start-up jitter.

Experiments on a silicon TRAPATT diode operating at 700 MHz with a power output of 70 W continuous were performed. Illumination was provided by a GaAs laser diode operating at 904 nm. Without illumination, the TRAPATT displayed about 100 ns of start-up jitter. When the TRAPATT was illuminated by a laser pulse positioned in time at the leading edge of the TRAPATT bias pulse, reduction in jitter occurred. Reduction to as low as 30 ns was observed with a laser output of 0.2 W/cm^2. This reduction in jitter was also made less temperature dependent.

2.2.4.3 Frequency Shifting

Experimental results on a silicon TRAPATT diode will now be discussed. Optical control of the frequency was noticed with laser illumination at levels greater than 10^{-2} W/cm^2 from a GaAs laser operating at 904 nm. The TRAPATT, whose frequency range was from 654 to 1493 MHz, could be shifted in frequency by about 1% at the low end of the band and by about 7% at the highest frequency when stimulated by the laser. In most cases, as the illumination level increased, so did the TRAPATT frequency.

2.2.4.4 Variation of Output Power

Using the same setup as in the previous discussion on frequency shifting, the power level of the TRAPATT could be controlled. The effect of the illumination depends on the frequency of oscillation of the TRAPATT. At the low end of the frequency band, the power lever decreases with increasing illumination, whereas at the higher frequencies, the power level increases with increasing power level. Power variation of about 9 dB was achieved when the illumination varied from 10^{-2} to 1000 W/cm^2.

2.2.5 MESFET Oscillator

An experiment on a MESFET oscillator was carried out to determine the capability to optically injection lock the oscillator [34]. This experiment was spurred by previous work on MESFETs which showed that the oscillator frequency could be tuned by varying the illumination intensity striking the active region of the MESFET. The illumination creates free carriers in the channel region that alters the DC characteristics of the device.

The MESFET was controlled by a GaAs/GaAlAs laser that produced an output power of 1.5 mW at 850 nm. The laser was modulated by a klystron oscillator that could be tuned around the MESFET oscillator frequency at about 2.35 GHz. Comparisons of the FM noise level of the free-running oscillator and the oscillator locked to the klystron were made. Measured in a 1 Hz bandwidth, the RMS FM noise deviation was only a few hertz at 10 kHz offset from the carrier. This is a significant reduction in the noise spectrum of the oscillator.

Locking range of the MESFET oscillator was also examined. A maximum locking range of 5 MHz was obtained. A greater locking range could be achieved if more efficient coupling existed between the laser and the FET chip, and also if more of the microwave signal could be coupled to the laser.

2.2.6 Transistor Oscillators

Transistors can be optically controlled by illuminating the base region with a light source. The light source generates carriers in the base that affect the conductivity of the transistor. In other words, the light source acts as an additional supplier of base current.

Experimental results on optically injection locking a transistor oscillator will be reviewed [35]. A silicon transistor with its lid partially removed to allow illumination was used. The illuminating device was GaAlAs laser operating at 820 nm. The laser output was intensity modulated by a signal generator with a frequency of 110 MHz. The frequency out of the transistor oscillator was 330 MHz which demonstrated that subharmonic optical injection locking took place. Fundamental frequency locking may also occur. The locking range of the oscillator was limited due to the same reasons discussed previously. First, the modulation method of the laser was inefficient; second, the coupling efficiency of the laser light into the transistor base region was low.

2.2.7 Optical Control of Passive Devices: Dielectric Resonator Oscillator

The DRO is used more readily as its benefits are realized. Some of the advantages of the DRO over other oscillators are reduced phase noise, higher efficiency, and lower susceptibility to microphonics. Figure 2.15 shows a simple depiction of a DRO circuit. Notice that the dielectric resonator is placed in close proximity to the output stripline that affects the feedback characteristics of the FET oscillator. The resonator couples magnetically to the feedback path and the distance between the resonator and the microstrip line determines the amount of coupling.

Optical control can be achieved by adding a photoconductive sample on top of the dielectric resonator material. Once illuminated, the conductivity of the photoconductor increases that alters the magnetic coupling of the dielectric resonator resulting in a frequency shift of the DRO.

2.2.7.1 *Illumination Effects on Dielectric Resonator Oscillator*

Two effects of illumination on DROs are discussed. First, frequency tuning, or shifting, is reviewed as with the IMPATT and TRAPATT diodes. Also, FM modulation of the DRO is discussed. Experimental results on other illumination induced effects are still in development.

2.2.7.2 *Experimental Results: Optical Tuning*

Experiments were performed on a 10.2 GHz DRO. Illumination was provided by three sources: white light, a HeNe laser with a power output of 5 mW at 630 nm, and a GaAs LED with a power output of 1 mW at 850 nm. The white

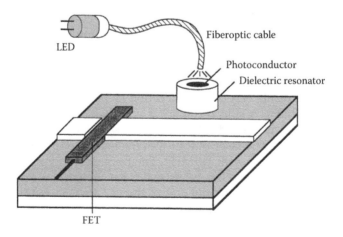

FIGURE 2.15
Optical tuning of the dielectric resonator.

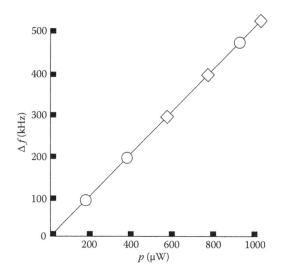

FIGURE 2.16
Shift in the DRO resonant frequency as a function of IA LED output power.

light source shifted the DRO center frequency by 5 MHz and also lowered the power by 1 dB. A total of 1 W/cm² of light output power was used. A 0.4 MHz frequency shift was obtained using the HeNe laser, and a 0.5 MHz shift was seen with the GaAs LED. In addition to a constant intensity illumination, the light sources were varied in intensity to determine their effect on frequency shift. Results showed a linear relationship between frequency shift and light intensity as shown in Figure 2.16 [36].

2.2.7.3 FM Modulation

FM modulation experiments were performed using an AM-modulated LED as the coupling mechanism to the photosensitive sample. Modulation rates of 130 MHz were observed; however, higher rates could have been achieved if the LED would not have limited the rate.

2.2.8 Applications of Optical Control

Applications of devices controlled by optical means are diverse and constantly increasing. A description of some common applications provides a better understanding of why much of the research has taken place.

One very important area for utilization optically controlled devices is in phased array radar systems. Phased arrays radars have many emitting elements (sometimes >10,000) on a planar surface. These emitting elements allow the antenna pattern to be electrically steered very quickly but, in order for the pattern to be correct, the phase and frequency of each one of the elements must be controlled rather precisely. In the past, these elements, which are actually oscillators, were controlled electrically either by coaxial cable or by waveguide. Because of the weight, volume, and expense involved in running >10,000 waveguides to a phased array radar, many radars that could have been phased array in the past were implemented in some other fashion. With optically controlled oscillators, however, these radars can now be implemented as phase arrays, because the control signal to each oscillator can be fed through small, lightweight, and inexpensive fiber-optic cable. Thousands of oscillators can be injection locked by a single master oscillator whose output is modulating a laser. Megawatts of output power can be achieved if IMPATT or TRAPATT diode oscillators are used. Millimeter wave operation is quite feasible.

Another radar application using optically controlled devices is in Doppler type radars. Depending on the frequency of radar operation and the velocity of the expected targets, the Doppler return could be very close to carrier frequency. Doppler offsets of 1 Hz are commonplace. In order for the Doppler to be detected, the system must be coherent in frequency and phase, and the noise level close to the carrier must be relatively low. Using the optical technique described in the IMPATT section, one could optically injection lock several oscillators in a system that would maintain frequency and reduce phase noise.

High-resolution radar applications could benefit from optically controlled devices. With the shorter pulse durations and the reduction in leading edge jitter (start-up jitter) afforded by optical control, the ambiguity in target range is reduced dramatically.

Frequency agile communications and radar systems are also candidates for optically controlled devices. The faster a system can "hop" from one frequency to another, the better. Subnanosecond frequency switching times exhibited by the optically controlled IMPATT diode oscillator mentioned earlier would be ideal.

Systems requiring immunity to EMI could use optical technology. Shipboard systems, for example, have to withstand extreme EMI environments because of the radar and communication gear on board. Since the optical signal would run through fiber-optic cable, the control signal would be inherently immune to outside EMI. Also, if there is information being transmitted through the cable that must be protected, the information is more secure than with electrical conductors.

Still another application is in airborne or portable systems where size and weight must be kept to a minimum. Using fiber cable reduces both of these critical parameters and also allows more complicated systems not possible without optical control. Finally, laboratory test setups are in need of devices and systems that are optically controlled. Laboratory work usually involves more precise control of parameters than in fielded systems. Narrower pulses and faster tuning can be achieved with optical control.

2.2.9 Future Needs and Trends

The future requirements of electronic systems warranting the use of optically controlled devices are many. Optical control promises to solve many of the current limitations of electronic systems. One of these is the need for faster switching devices. Switching speed improvements are required for on/off, frequency, and power level switching. Wider bandwidth systems are also needed to improve the locking range of injection locked systems. Improving the locking system will allow even more modulation of the carrier frequencies. With this comes wider bandwidth capability and the requirement for higher modulation rates.

Another requirement will be to control electronics with less optical signal power. Currently, the coupling efficiency between the light source and the microwave device is very poor. Higher reliability and producibility requirements will follow the successful laboratory demonstrations of new concepts. Size and weight reduction will be stressed.

One key trend is the fabrication of optical and microwave components on a single monolithic device. As discussed in Section 5.11, much of the MMIC technology can be directly applied to the development of optics modules. Optomicrowave monolithic integrated circuits will assist in obtaining better control of the microwave devices with less optical power because the coupling efficiency will increase. Size, weight, reliability, and manufacturability will ultimately improve with the OMMIC systems as well. Laser technology will also continue to improve on the current limitations of optical control. Higher modulation rates and noise reduction improvement will parallel the improvements in laser technology.

2.3 Lithium Niobate Devices

A large number of important devices that are fabricated in $LiNbO_3$ have been discussed in the literature. Some of the most interesting of these are described in this section [37].

2.3.1 Optical Switches

One of the most important $LiNbO_3$ devices is the optical switching device. Response times as fast as 1 ps are possible [38] due to the electro-optic effect inherent in lithium niobate, so that high-speed switching is limited merely by the circuit delays that result from the finite length of these devices. The fundamental building block of any optical switching network is the simple changeover point or 2×2 optical cross point shown in Figure 2.17 [39]. In the through state ports, 1 and 3 and 2 and 4 are connected. In the crossed

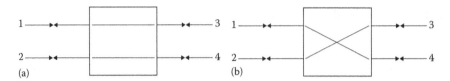

FIGURE 2.17
(a) Simple changeover point and (b) 2 × 2 optical cross point.

state, 1 is connected to 4 and 2 is connected to 3. This crossed state may be achieved by mechanical means such as by physically moving the fiber or a prism into the optical path. If the waveguide is made from a material such as LiNbO$_3$, however, then the crossing may be caused by applying a voltage and utilizing the electro-optic effect. The latter is obviously preferred due to difficulties in aligning microfibers, prisms, etc., in a mechanical setup. Currently, typical LiNbO$_3$ switches have insertion losses of 3–5 dB, extinction ratios of 20–30 dB, and require control voltages of 2–30 V depending on switching speed. For mechanical microoptic devices, these values would be 1.5 dB, >60 dB, and 14 V, respectively [40]. The cost of each type of device is roughly the same; however, LiNbO$_3$ switches are rapidly evolving, and improved device performance as well as greater reliability and lower cost will push the balance in favor of that technology.

2.3.2 Directional Couplers

The basic directional coupler shown in Figure 2.18 [41] consists of two identical single-mode channel waveguides fabricated in close proximity via Ti diffusion over a length L. This allows synchronous coupling between the overlapping evanescent mode tails. Typical values of L are on the order of 1–10 mm. If the velocity of propagation is the same in both guides (same

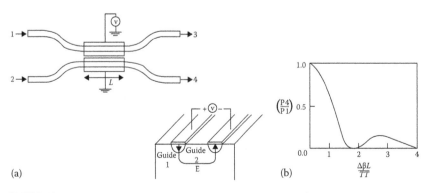

FIGURE 2.18
(a) Directional coupler and (b) transfer characteristic.

refractive index), then the light will be entirely coupled into the second guide after a characteristic length referred to as the coupling length. This coupling length value is definable in terms of wavelength, mode confinement, and interwaveguide separation [42]. If the indices of refraction of the two guides are slightly different, the net coupling may be quite small, so that application of a voltage across the electrodes shown in Figure 2.18 will change the indices of refraction and, therefore, the propagation velocities, thus destroying the phase matching. This determines which port (3 or 4 in Figure 2.18) the light will be coupled into after a length L. For directional couplers, switching speeds as fast as 1 ps are possible; however, to accomplish this device, lengths must be on the order of 1 mm. Although not a fabrication problem it is known that the shorter the device length, the greater the applied voltage must be in order to obtain optical isolation [43]. For device lengths of 10 mm, control voltages of 2–8 V are required dependent on the wavelength. For 1 mm devices, voltages of 10–30 V are required. Two other examples of directional couplers are the X switch and the merged directional coupler (Figures 2.19 and 2.20) [44]. Again a applied voltage will cause a change in refractive index to occur causing changes in the coupling ratio. Switching characteristics are direct functions of device length and switching voltage as with previous devices.

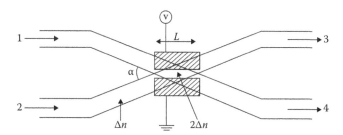

FIGURE 2.19
X-switch.

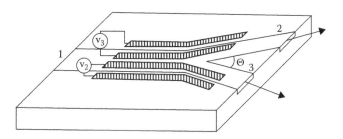

FIGURE 2.20
Y junction switch.

2.3.3 Modulators

Two important interference type waveguide modulators are Crossed-Nichol and the Mach Zehnder devices shown in Figures 2.21 and 2.22 [45]. In the Crossed-Nichol architecture, metal electrodes are formed on the crystal surface adjacent to the waveguide. The input beam is then divided into two components (TE and TM propagating mode) and due to the material birefringence, the separate components will be phase modulated differently as a voltage is applied to the electrodes. In the Mach Zehnder type, the input beam is divided into two identical interferometer arms. These arms may be thought of as two phase modulators and the applied voltage may be adjusted, so that the signals are either in or out of phase on recombination. The voltage required for a phase shift is [46]

$$V_x = \frac{\lambda g}{2\delta n_e^3 r_{33} L_e} \tag{2.3}$$

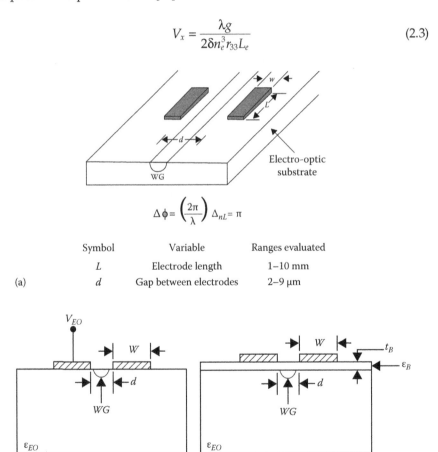

$$\Delta\phi = \left(\frac{2\pi}{\lambda}\right) \Delta_{nL} = \pi$$

Symbol	Variable	Ranges evaluated
L	Electrode length	1–10 mm
d	Gap between electrodes	2–9 μm

(a)

(b)

FIGURE 2.21
(a) The guided-wave phase modulator layout and (b) basic guided-wave phase modulator surface electrode architectures.

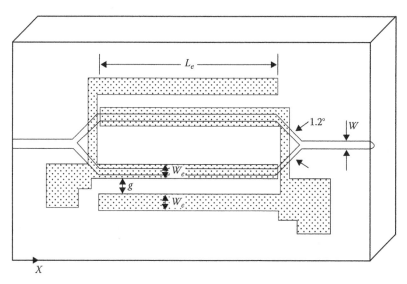

FIGURE 2.22
Schematic representation of modulator structure.

where
    λ is the wavelength
    g is the gap between the electrodes
    δ is the electric/optical field overlap factor
    n_e is the extraordinary refractive index
    r_{33} is the material electro-optic coefficient
    L_e is the length

If out of phase, the signal no longer propagates along the waveguide and radiates away into the LiNbO$_3$ crystal. If in phase, the light remains in the guided mode and continues to be transmitted through the device. The intensity-modulated outputs are

$$l = l_0 \sin^2\left(\frac{\Delta\phi}{2}\right) \text{Crossed - Nichol} \tag{2.4}$$

$$l = l_0 \cos^2\left(\frac{\Delta\phi}{2}\right) \text{Mach Zehnder} \tag{2.5}$$

where l_0 = input light intensity. Mach Zehnder type devices generally provide up to 3× the efficiency of the Crossed-Nichol devices.

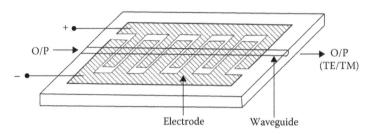

FIGURE 2.23
Polarization TE-TM modulator/converter.

2.3.4 Polarization Controllers

Another important device is the polarization controller (Figure 2.23) [47]. In these devices, two interdigitized electrode fingers are formed over a single waveguide. The electrodes affect the two components of the general elliptical polarization state (TE and TM) modes by modifying their relative amplitudes and phases. The voltage in the interdigitized section of the electrode changes the relative polarization amplitudes of the modes, while the voltage in the uniform portion acts to change the relative phases of the modes. Effective coupling between these orthogonal modes may be achieved with conversion efficiencies better than 99% [48], with an applied voltage of 13 V and an electrode period of 7 μm. The phase matching condition is [49]

$$2\left(\frac{\pi}{\lambda}\right)(N_{TE} - N_{TM}) = \frac{2\pi}{\Lambda} \tag{2.6}$$

where
  Λ is the electrode period
  N_{TE} and N_{TM} are the effective waveguide indices

2.3.5 Integrated Systems

Several other device types exist such as Bragg effect modulators (Figure 2.24), frequency translators [50], and tunable wavelength selectors. These all operate in similar fashion to those devices already described. Although any of these devices may be separately fabricated and then assembled into packages via single-mode splices or connectors, this approach is simply not practical. This is because the insertion losses and the system costs would be excessive, especially for systems requiring a large number of components. These systems can be made practical via the integration of several components onto a single substrate. Examples of such integrated devices that have been fabricated include: a coherent optical receiver (Figure 2.25) [51] and directional couplers (Figure 2.26) [52]. The only limit to integration arises from the

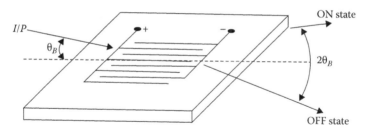

FIGURE 2.24
Bragg-effect electro-optic modulator.

required device lengths since the cell sizes for integrated optical devices are quite large relative to electrical integrated circuits. Device lengths of several millimeters are required if high modulation rates are to be achieved. Device widths are less, however, and there is a great potential for arrays of identical devices to be fabricated onto a single crystal of $LiNbO_3$ with parallel inputs. At this time, the best lightwave system performance of 8 Gbs/s over an unrepeated distance of 68 km [53] is achieved via the use of $Ti:LiNbO_3$ devices. As the technology matures, these devices are expected to play a critical role in future optical signal communications.

2.4 Applications of Fiber-Optic Systems

Fiber-optic transmission is utilized in data communications and telecommunications for local area networks and telephone trunks. Video transmission, closed circuit television, cable television, and electronic news gathering utilize fiber optics, while in the military, command, control, and communications and tactical field communication systems are being developed and installed with fiber links. Optical circuits are applicable in each of these applications as interconnecting components in the system. Active devices such as sources and detectors are utilized in transmitters and receivers, and a number of passive devices perform various critical functions in the fiber-optic system. These functions include switching and routing, modulation, regeneration, processing, sensing, and multiplexing. Devices include taps, distributors, couplers, concentrators, switches, relays, multiplexers (MUXs), cross connection cabinets, and gratings. Local area network topologies currently in use as well as those in the development stages require a number of devices for the most efficient operation. A growing trend is toward all-optical switching systems, eliminating the use of slower electrical routing of information.

When comparing the performance of fiber transmission media to the alternative electrical twisted pair or coaxial cable conductor, the ability to utilize

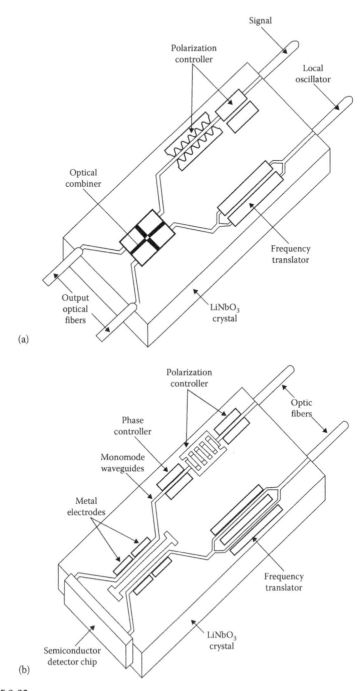

FIGURE 2.25
(a) Z-cut LiNbO$_3$-integrated optic coherent receiver device. (b) X-cut LiNbO$_3$-integrated receiver device.

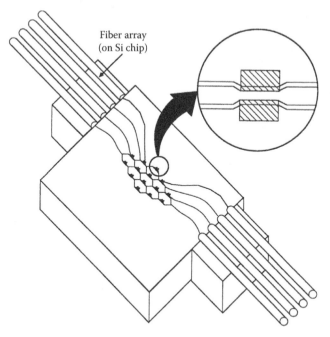

FIGURE 2.26
Diagram of a 4 × 4 directional coupler switch matrix.

the wider bandwidth of the fiber without an increase in attenuation is a clear advantage. In addition, significant size reductions and greater resistance to long-term environmental degradation provide advantages to the fiber medium.

2.5 Optical Interconnects for Large-Scale Integrated Circuits and Fiber Transmission Systems

2.5.1 Introduction

The limitations of conventional interconnects and switching technology are rapidly becoming critical issues in the throughput of data within the high-speed signal processors using LSI chips or GaAs-integrated circuits. The problems associated with transmission of Gb/s data must be addressed when designing integrated systems utilizing electrical interconnects and conventional switching technology. The performance of electrical interconnects at high data rates is adversely affected by increases in capacitance and reflections due to impedance mismatches. Optical interconnect technology promises to significantly enhance signal processing systems and provide relief from pinout, physical proximity, and clocking problems [54]. Furthermore, by releasing the bandwidth constraints imposed by electrical interconnects, the full processing capability of LSI chips could be exploited to improve currently fielded systems. Practical interconnection at the intra-board, backplane, and cabinet levels of signal processor system can now be realized with optical interconnects.

Faster logic switching times and increased use of very large-scale integration technology are placing new demands on digital system interconnections. In particular, board and backplane interconnect designs for future systems must satisfy requirements for increased density and improve electrical performance with respect to transmission of high-frequency signals. System parameters such as rise time, impedance level, and the assignment of ground return paths will have a significant effect on the high-frequency performance of electrical interconnections.

In order to realize the potential of the new families of fast switching components, there is an increasing need for system electrical interconnections to be designed as networks of transmission lines. Logic switching times continue to decrease at a much faster rate than the length of typical interconnections on the board or backplane level [55].

Interconnects may be classified by three categories according to their system-level utilization. These categories may be termed intraboard, backplane, and cabinet levels. The primary distinguishing characteristic of these categories is the interconnect length. Definitions of these lengths, albeit somewhat arbitrary ones, can be made. Intraboard level interconnects are approximately 5–15 cm long. Backplane level interconnects may run from 15 to 40 cm, while the longest interconnects are at the cabinet level and are typically on the order of meters.

2.5.2 Link Design and Packaging

The very large number of signal connections that must be made to densely integrated chips creates a very high density of separate conductors on the printed circuit boards. Multilayer boards relieve coupling and crossover problems by allowing crossovers to occur in different layers. These boards may be assembled as multilayered, multichip carriers that also contain metalized "vias" for vertical signal conduction. The number of chips and the dimensional extent of the modules is limited by the capacitance load that the chip is capable of driving without excess deterioration of rise time and/ or propagation delay. The use of special drivers or terminated transmission lines is not anticipated for these short internal connections.

One the other hand, the multilayer board structure does allow signal paths to be implemented in an approximation to the balanced stripline or the microstrip geometry; therefore, multilayer boards will permit terminated lines for point to point connections. Interspersal of signal lines between ground and power planes reduces crosstalk. Additional "shield" layers may be introduced while clock distribution on the boards can be accomplished with equal phase length-terminated branches. Various practical empirical rules have been determined for similar particular semiconductor technologies such as TTL compatible CMOS [56].

2.5.3 Backplane Interconnects

Backplane connections are proportionately long and the mutual inductive and capacitive coupling between conductors produce unwanted and intolerable crosstalk between circuits carrying fast rising signal waveforms. Backplane electrical design centers around minimizing and controlling these couplings. It is important that the couplings remain the same in successive copies of systems and not change during normal operation or servicing. To this end, orderly arrangement of printed circuit panels, ribbon cables, twisted pairs, and coaxial cables are used for these connections. Special interfacing line drivers are required for high-speed lines between chip modules.

Normally, a backplane structure is passive, serving only to supply interconnection and mechanical support. In some cases, active circuitry has been incorporated in the backplane presumably to reduce the lengths of some connections or to equalize lengths. This arrangement can create serious troubleshooting and maintenance problems however; and some systems made in this way are being retrofitted with passive backplanes.

2.5.4 Power Distribution

In multilayer modules, power distribution is accomplished by adding an additional layer to act as a power bus. By employing wide traces with minimum lengths, series inductance is minimized. In this manner, power busses can be considered as a line with low characteristic impedance.

A phenomenon that occurs in RF circuits and is detrimental in very fast digital circuits is the creation of relatively high Q resonators when the length dimension of the lines approaches a quarter or half wavelength of signal in the dielectric substrate. In some cases, the Q can be "spoiled" by introducing resistance across the bus at the ends.

Power distribution is a key determinant of a backplane's performance and reliability. It is usually accomplished by heavy vertical bus bars. Before choosing a backplane configuration, the system's current flows should be understood. Proper placement of inputs and returns ensures current densities will be uniform and voltage drops low enough to permit reliable operation. Paralleling connector pins at power entry points requires a significant derating of the current handling capability of the pins. Any small difference in contact resistance will significantly unbalance the current flows.

2.5.5 Large-Scale Integration Challenges

Many problems and issues remain to be resolved in the packaging and general usage of large-scale integration chips. The first of these issues concerns high-level interconnections among LSI components and devices. Although the development of LSI technology has been rapidly progressing, the only major interconnection issue that has been addressed has been at the chip-to-chip level. It is felt that a critical path in the incorporation of LSI components into various designs will be interfacing at the backplane and intercabinet levels. This will be particularly true for the interface speeds of LSI components.

The second issue is the actual insertion of LSI components into existing systems and the incorporation of these components into new product designs. The new problems of LSI electrical interconnects stem from two sources. First is the necessity to transmit high-speed data and clock signals over distances that are incompatible with the drive capabilities of the chips. Second is the lack of control of clock offsets between widely separated functional entities.

A prime consideration affecting communication between chips is the drive capabilities of the chips themselves. Additionally, these limitations do not allow the direct driving of any currently available optical sources. Analysis suggests that the critical length separating lumped element and transmission line considerations is on the order of 10 cm. This was obtained from calculations using a rise time of 7 ns as representative of typical 25 MHz chip signals.

A hierarchical approach to system implementation is envisioned wherein chips containing closely interacting functions or groups of functions are assembled in close spatial proximity as higher function modules. Combinations of superchip modules and line drivers should suffice for structuring circuits at the board level. The design goal is to minimize the number and length of high-speed transmission paths. Experience indicates the number of interconnections between assemblies decreases as one moves up through higher level assemblies. In the case of fewer high-speed channels traversing greater distances, higher power transmitters and more sophisticated receivers can be tolerated. Also, the requirement for synchronous clocking diminishes at higher assembly levels. At certain levels in signal processors, we move from intra- to intercomputer situations. These widely separated elements need to communicate over high-speed data connections—data connections that are impracticable given the chip drive capabilities.

2.5.6 Advantages of Optical Interconnects

There are a number of limitations of conventional interconnects that can be alleviated through the use of optical interconnects. The general advantage of optical interconnects over their electrical counterparts are the following:

- Freedom from stray capacitance and impedance matching
- Freedom from grounding problems
- Provide relief from the pinout problem
- Lower power requirements
- Increased flexibility for interconnects
- Light weight and small volume
- Planar integration
- Increased effective bandwidth of the system
- Two-way communication over a single transmission path (fiber)
- Passive MUX/DEMUX for high reliability at low cost
- Major system cost reductions
- Simple upgrading of existing systems
- Immunity to RFI, EMI, and EMP effects

The first, and perhaps most important, advantage is the lack of stray capacitance. Optical interconnects suffer no such capacitance effects and crosstalk can be controlled as long as care is taken to avoid scattering effects in broadcast systems. The second advantage is freedom from capacitance loading effects. The speed of propagation of an electrical signal on a transmission line depends on the capacitance per unit length. As more and more components are attached to a transmission line, the time required to charge the line increases and the propagation speed of the signal decreases. Optical interconnects provide a constant signal speed (i.e., the speed of light in the guiding medium), which is independent of the attached components.

2.5.7 Compatible Source Technology

The telecommunications and optoelectronic industries have pursued extensive research programs investigating semiconductor optical sources operating at 0.85, 1.3, and 1.55 μm wavelengths. This has introduced to the marketplace a wide range of available sources that are capable of achieving previously unattainable modulation rates with extremely high reliability. LEDs are generally limited to hundreds of Mb/s up to 500 MHz [59]. In contrast, laser diodes, although similar in structure to LEDs, are highly efficient. An experimental InGaAsP laser has demonstrated an internal quantum efficiency approaching 100%. This cooled laser has also shown a bandwidth capability exceeding 26 GHz [60]. Depending on the system implementation used in the interconnection of LSI components, either of these types of sources would be adequate for data transmission.

High bit rate communication sources at high-speed pulse code modulation rates should have the following:

- No modulation distortions (pattern effects)
- Narrow spectral bandwidth
- No high spectral broadening due to modulation
- No self-pulsation (charge storage effects)

Currently at issue are the power requirements for both types of sources that range from 7 to 100 mA for adequate output in LEDs or threshold operation in laser diodes. The critical future laser requirements for the optimal usage of optical interconnections are "development of continuous wave room temperature frequency-selective lasers with high stability (10 Å) and low threshold current (5 ma)" [61] that are capable of being modulated in the tens of gigahertz just at or above threshold.

2.5.8 Receiver and Detector Technology

Optical interconnect receivers are basically high-speed optoelectronic transducers that receive incoming optical digital data, transform it to an electrical signal, amplify and filter it, and restore logic levels. There are several alternatives for detectors. Among these are p-i-n photodiodes, avalanche diodes, and Schottky barrier photodiodes. The p-i-n photodector is essentially a p-n junction with a controllable depletion region width—the "i" intrinsic region. Incoming photons generate electron–hole pairs in the intrinsic region and the carriers are swept out of the region by an applied electric field. The device speed is chiefly determined by the drift mobility. As a result, the fastest response times are achieved using GaAs devices. The voltage requirements to deplete the active region are on the order of 5–10 V.

The APD is also a p-n diode; however, it operates in a highly biased mode near avalanche breakdown. Its operation is similar to the operation of a photomultiplier tube with improved sensitivity of 3–10 dB. APD detectors can exhibit rise times in the 30–40 ps range and 80%–90% efficiency [62]. Schottky barrier photodiodes possess the advantage of being ideally suited for surface detection. They are constructed by using a thin, transparent layer of metal for one contact that allows the light to pass very efficiently into the semiconductor contact. With proper construction, any of these detectors are capable of detecting data rates in the gigahertz range and are compatible with the wavelengths and modulation rates of laser diodes and LEDs.

2.5.9 Integration of Sources and Detectors

Due to the telecommunications industry requirements, a large number of sources and detectors are available in single-device packages from commercial vendors. Space and power limitations become unacceptable, however, when considering these devices for use in smaller and faster computing environments. Such environments necessitate the development of lower power and high-density devices and packages [63]. In particular, packaging methods must be developed to allow high-speed silicon and GaAs components to be integrated with optoelectronic components and waveguides [64]. Direct integration of optoelectronic devices with conventional logic devices will increase density and reliability of the interconnect because the electrical–optical interface occurs on chip [65], eliminating the need for hybrid or separate packaging techniques for the optics.

If a photodetector is to be integrated on a circuit with more than a few electronic components, it must meet a number of criteria [66]. First, the detector must be compatible with electronics processing. It must therefore be processed on a production line and be compatible with the substrates used for the electronics. For GaAs, this means at least a 3 in. semi-insulating substrate. Second, the material for the detector cannot interfere with the electronics. Thus, if epitaxial material is required, it must be excluded from the regions in which the electronics will be fabricated and the transition to the epitaxial region must be smooth enough to permit fine-line photolithography. Third, any process needed to fabricate the detector cannot degrade the performance of the electronics. For example, a very high temperature step may cause unacceptable surface damage. Fourth, the detector and electronics must be adequately isolated on the substrate.

Finally, the integrated photodetector must meet the overall receiver system specifications. The receiver's function is to convert the optical signal to an electrical signal compatible with digital electronics. This requires coupling the input fiber to the detector, designing and fabricating a detector with sufficient bandwidth and sensitivity, interfacing the detector with a preamplifier, and converting the analog signal from the preamplifier to a digital signal.

The sensitivity of the receiver depends strongly on the node capacitance, with the improvement being most significant for a reduction in capacitance at the highest bit rates. For example, a reduction in mean detectable optical power of nearly 5 dB is possible if the front-end capacitance is reduced from a relatively good hybrid receiver value of 1 F to a value of 0.2 pF at $B = 1$ Gb/s. This additional margin might then be used to permit a less-expensive coupling scheme or to power split from a laser to multiple detectors in an optical bus. Alternatively, the fivefold decrease in capacitance would permit the detector amplifier to be operated at nearly five times the bit rate with no degradation in accuracy. This is a strong argument for integration.

While the integration of the detector with the amplifier may prove to be the limiting factor on the speed of the optical link due to circuit limitations, the integration of the laser with associated electronics will be more difficult from a materials and processing compatibility standpoint. There are three reasons for this. Lasers require multilayered heterostructures up to 7 μm thick. They also need two parallel mirrors separated by on the order of 200 μm, and finally, a method for achieving electrical and optical confinement in the lateral dimension.

2.5.10 Integrated Device Developments

There have been a number of demonstrations of laser diodes integrated with either a single or a few GaAs IC components [67]. For the past several years, a number of companies, including Honeywell and Rockwell, have been developing technology to integrate optoelectronic devices with complex GaAs circuits [68]. One structure that has been demonstrated [69] is shown in Figure 2.27. The components in the structure are a transverse junction stripe (TJS) laser, a FET driver, and a 4:1 MUX. The MUX and driver active regions are formed by selective ion implantation into the semi-insulating GaAs substrate, while the TJS laser is fabricated by liquid phase epitaxial growth in a well etched into the substrate. The rear laser mirror facet is formed with a microcleave process developed at the California Institute of Technology, CA [70]. A cross-sectional view of the microcleave process is shown in Figure 2.28. The small horseshoe-shaped wing at the end of the laser is formed by chemically etching under the Al-facet. The wing is broken off to form a smooth cleaved mirror facet. Measurements have shown that lasers with undercut mirrors have operating characteristics (laser threshold 37 mA) identical to similar

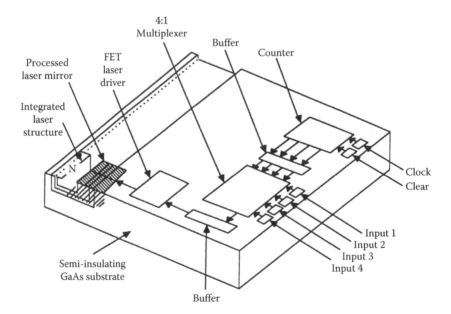

FIGURE 2.27
Block diagram of the monolithic optoelectronic transmitter consisting of an integratable TJS Laser, a laser driver, and a 4:1 MUX.

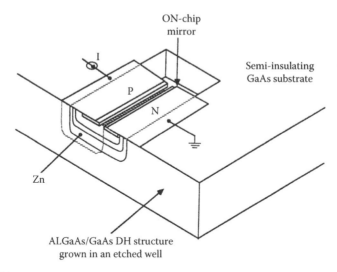

FIGURE 2.28
Schematic representation of the integratable TJS laser grown in an etched well. The on-chip mirror facet is formed by the undercut mirror process.

lasers having two cleaved mirrors. The optoelectronic chip having a size of 1.8 mm × 1.8 mm operated at speeds up to 150 MHz [71].

More recently, a 1 Gb/s optoelectronic receiver chip was demonstrated [72]. The receiver chip was designed to digitally multiplex four high-speed input signals. Two of the inputs are electrical and two are optical. The receiver consisted of two back-to-back Schottky diode detectors, two amplifiers, a 4:1 MUX, and a laser driver. All of the components were fabricated on a semi-insulating GaAs substrate using direct implantation MESFET technology.

The photodetector was fabricated on a semi-insulating substrate, thus eliminating the need for epitaxial growth. The pulse response of the back-to-back Schottky photodiodes [73] had rise and fall times for a range of bias voltages between −10 and −15 V of under 100 ps, which is sufficient for a 1 Gb/s operation of the receiver. The responsivity of the detector was measured to be 2.0 A/W for a wavelength of 0.84 μm. The output of the detector is fed into a three-stage amplifier, a preamplifier, a gain stage, and a buffer stage. The preamplifier translates the current into a voltage and amplifies to the proper GaAs logic levels by the gain stage, and the buffer stage is used as a line driver for the input of the 4:1 MUX. The preamplifier had a gain of 20 dB with a feedback around the second stage to improve its frequency response.

A similar effort on monolithic integration using a multiple quantum well (MQW) laser with a ridge channel waveguide structure [74] is shown in Figure 2.29. The MQW laser structure was grown by MOCVD and consisted of five 100 GaAs wells separated by four 40 Å $Al_{0.2}Ga_{0.8}As$ barriers. The ridge waveguide was ion milled having a width of 5 μm. These integrated devices

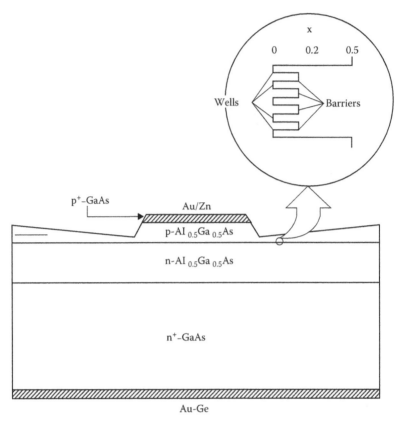

FIGURE 2.29
Cross-section and energy band diagram for ridge waveguide MQW laser.

have shown room temperature operation with 15 mA thresholds and differential quantum efficiencies of 60%. For this particular integration scheme, the laser will be implemented as a surface emitter as shown in Figure 2.30. The laser light is coupled from the laser to a passive AlGaAs waveguide. The light in the waveguide is coupled from the surface either with an etched V-groove or a grating. This technique allows the user to fabricate lasers at any position on the chip rather than only at the edge of the chip. The method is more difficult to implement if fibers are to be used as the transmission medium; however, it does provide the flexibility for utilizing the third dimension and directing the optical signals via holographic techniques. The photodetector used in this development was a PIN structure grown by MOCVD. The device was a six-layer stack consisting of GaAs and AlGaAs layers. Initial tests on this device [75] have shown a dark current of <10 pA and a breakdown voltage of 80 V. A current gain of 10 was demonstrated at a voltage of 95% of breakdown.

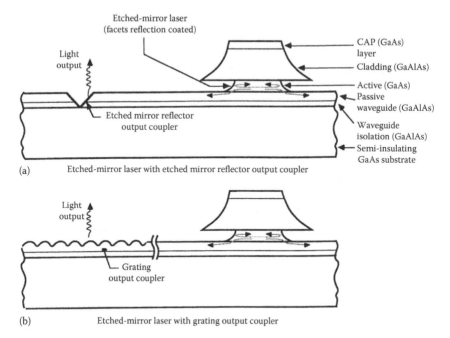

FIGURE 2.30
Surface-emitting laser using (a) etched V-groove and (b) grating.

2.6 Optical Interconnect Media

Optical interconnection of LSI chips requires a passive distribution network to transmit optical information from one point to another. To permit the general interconnection of one transmission point with other reception points in the system, the data are transmitted in a data packet from the various transmitters. As a result, the information is contained in a burst of bits, some of which indicate the destination, others the sender, and others the data to be transmitted. The nature of the transmission can be point to point or global. As a result, it is required that the optical interconnect medium permits efficient distribution of optical energy to all the required ports, with a minimal loss due to coupling and insertion loss of the optical signal. Furthermore, the dimensions of this optical medium must be compatible with the sizes of typical integrated circuits and systems utilizing such circuits.

2.6.1 Guided Wave Interconnects

The guiding of light for transmission via optical fibers is accomplished by confining the optical energy in a core region that is surrounded by a cladding layer of lesser refractive index. This refractive index profile from the core to the cladding generally takes two forms—either a sharp step function change (step-index) or a sharp change followed by a graded change (graded-index) in the core.

The two key differences between single- and multimode fibers are the dispersion characteristics and the dimensions. Dispersion causes signal degradation and therefore limits the distance over which transmission can be effectively accomplished. The dimensions of the guiding cores differ greatly between the two types of fibers causing ease of coupling and effectiveness of alignment to become issues.

Thin-film planar waveguides confine light using the same principle of differing refractive indices. In principle, the practical determining factors as to which materials can be used for waveguide fabrication are the refractive index, loss coefficient, and available dimensions. It is the last factor that promotes the use of thin films for the guides. For many acceptable materials, the refractive indices are such that the required waveguide dimensions are on the order of microns. Thin-film processing has advanced to the stage where films of this thickness are easily obtainable. Of special interest are waveguides in semiconductor materials. GaAs in particular has become increasingly attractive because of the compatibility with monolithic electronic and electro-optic device fabrication.

2.6.2 Single-Mode versus Multimode Fibers

Single-mode step index fibers possess the ultimate bandwidth capability due primarily to favorable dispersion characteristics. This means that single-mode fibers are the optimal choice for long haul applications in data transmission. In some applications, however, the lengths of transmission are such that multimode fiber can be utilized. A prime advantage of multimode fibers is the greater ease of coupling. Multimode fibers have acceptance areas of many orders of magnitude greater than single-mode fibers and thus have greater tolerance to misalignment and provide higher system gain.

Fiber dispersion can be classified as either intermodal or intramodal dispersion. Intermodal dispersion is found primarily in multimode fibers and is due to the differential delay between modes at a single frequency. This is eliminated in single-mode fiber since only one mode can propagate. Intramodal or chromatic dispersion is due to the variation of group velocity (the speed at which energy in a particular mode travels along the fiber) with wavelength.

For multimode graded-index fibers, the dominant cause of pulse spreading is intermodal dispersion. Experimentally determined values of this pulse dispersion have been found to range from 20 to 500 ps/km [76]. Dispersion values can be calculated exactly for quadratically graded index fibers. The results for calculations using an equation derived by Yariv [77] and particular fiber characteristics are shown in Figure 2.31. A reasonable bit rate for LSI applications would be 1.6 Gb/s. This assumes time division with sixty-four

f_{max} = Hz	No. Modes	Wavelength	Length
25 MHz	30	1 μm	4.98×10^3 m
100 MHz	30	1 μm	1.25×10^3 m
1.6 GHz	30	1 μm	77.9 m
6.4 GHz	30	1 μm	19.5 m

$$\Delta T = \frac{n \sqcup}{c} \left[\frac{n^2}{2nk^2} (l_{max} + M_{max} + 1)^2 \right]$$

$$f_{max} \simeq \frac{1}{\Delta T}$$

FIGURE 2.31
Quadratically graded index multimode fiber dispersion characteristics.

25 MHz lines. The results of Figure 2.31 and the values quoted indicate that multimode graded-index fiber would be adequate to transmit over distances of approximately 20 m. With the use of single-mode fibers, the pulse spreading is due to group velocity dispersion. This dispersion, in turn, is dependent on the spectral width of the source.

2.6.3 Broadcast Interconnects

The second technique for interconnecting various parts of high-speed integrated circuits involves the broadcast transmission of the optical signal in free space. Since substrates are planar, it can be envisioned that interconnection paths in free space must be perpendicular to the substrate plane to take advantage of the inherent three-dimensionality of free-space interconnects.

2.6.4 Free-Space Interconnects

With the free-space interconnect, the signal transmitted by the source is collimated by a lens or guided by a lens system. Detectors integrated into chips on the board receive the optical energy and convert it to electrical energy. The detectors can be made to receive the information with identical delays, due to the particular location of the source with respect to the focal point of the lens or mask. The system is highly inefficient, however, because only a small portion of the optical energy falls on the photosensitive areas of the chip. This method may therefore require additional amplification of the detected signal. This leads to increased power needs for boards, so that with many interconnects, this can be prohibitive in terms of heat dissipation and power requirements. Moreover, optical energy falling on areas of the chip where it is not wanted may induce stray electronic signals; therefore, it is necessary that opaque dielectric blockers be used. Openings in this layer would allow the detectors to receive the optical power.

2.6.5 Holographic Interconnects

Holographic interconnects fall into a class called free-space "focused" inter-connects, which can also be called "imaging" interconnects. For such inter-connections, the optical source is actually imaged by an optical element onto a multitude of detection sites simultaneously. The required optical element can be realized by means of a hologram that acts as a complex grating and lens to generate focused grating components at the desired locations. The efficiency of such a scheme can obviously exceed that of the unfocused case, provided the holographic optical elements have suitable efficiency. Using dichromated gelatin as a recoding material, efficiencies close to 100% can be achieved for a simple grating. The higher the number of focused spots, the lower the efficiency of available holographic methods. The flexibility of this method is excellent for any desired configuration of connections.

The chief disadvantage of the focused interconnect technique is the very high degree of alignment precision that must be established and maintained to assure that the focused spots are striking the appropriate locations on the chips. The spots can be intentionally defocused, with a trade between effi-ciency and alignment tolerance.

With holographic optical elements, a future possibility is the incorporation of dynamic holographic materials, such as those now being studied for four-wave mixing optical phase conjunction applications. Other future forms are banks of holographic mapping elements in conjunction with real-time masks. Candidates for such interconnections would be matrix-addressed liquid-crystal devices or the matrix-addressed spatial light modulator [78]. Another approach for a dynamic interconnect is the implementation of an optical matrix-vector multiplier [79].

2.6.6 Guided Wave versus Broadcast Interconnects

This represents the most critical trade-off decision as it determines the course of many others. Guided wave interconnects are inherently more efficient than broadcast interconnects. Since GaAs ICs will operate in the Gb/s range, the sensitivity of receivers is limited. As a result, the optical losses in the trans-mission of data from one board to another are limited to $-27\,dB$ for a signal-to-noise ratio of 100 (bit error rate $= 1 \times 10^{-10}$) [80]. Broadcast interconnection between devices on a board would require the use of large-area detectors and is not suitable for GaAs ICs. On the contrary, broadcast interconnects allow a large fanout of data, but guided wave interconnects also meet the needs of a large fanout with little or no loss occurring during transmission.

Optical fibers are currently the preferred choice. Owing to their flexibility, fibers can accommodate complicated topologies more readily than planar waveguides or broadcast methods. As mentioned, multimode fiber is more tolerant to misalignment than single-mode fibers, and is more tolerant than all other methods.

2.7 Multiplexing and Demultiplexing: Information Distribution Techniques: WDM Schemes

There are many advantages to wavelength division multiplexing (WDM) systems where lightwaves with different wavelengths, each carrying different information, are transmitted simultaneously through a single fiber. WDM data communication systems rely on the use of optical MUXs and DEMUXs designed to assure low insertion loss per wavelength channel, high isolation between channels, small size, and high reliability.

Several fundamental elements that have been considered for WDM applications are prisms, gratings, and interference filters, although other novel approaches exist. These components are the passive optical components that can perform the efficient multiplexing and demultiplexing of light sources into and out of trunk interconnections. In a data interconnect system, WDM is most advantageously utilized with optical fibers as opposed to other interconnection schemes.

2.7.1 Prisms

These MUX/DEMUX devices make use of the birefringement properties of various transparent materials. In general, this uses the principle of the dependence of optical speed in the material on wavelength. Two companies released representative product information on such devices [81]. GTE designed a two-channel system that can utilize source wavelengths that are 13–23 nm apart and have spectral widths of up to 6.5 nm. This is accomplished by using adjustable length polarizing prisms. The prism length is adjusted such that the two wavelengths emerge perpendicular to one another. ADC fiber optics offers both passive and active MUX/DEMUX devices. The passive device utilizes a biconical taper technique in which the coupling length of the device is wavelength dependent. Each wavelength, therefore, will exit at a different port of the device. The active device utilizes a dichroic filter to separate the different wavelengths. Although this type of filter is generally bulky, the ADC device is produced in a high-precision injection-molded plastic package that is small enough to be PC board compatible.

2.7.2 Gratings

In grating MUXs and DEMUXs, the mixture of wavelengths arriving in the transmission fiber or from the laser is separated and passed on to the receiver fibers. The light emerging from the transmission fiber or laser can be collimated by lenses before striking the grating. The angle at which the light is diffracted by the grating in relation to the optical axis depends on the wavelength. Several different gratings are currently available that have different diffraction efficiencies. One such MUX/DEMUX uses a chirped grating. This is an effective means of demultiplexing several channels, although gratings are somewhat bulky in comparison to integrated structures.

2.7.3 Bandpass Filters

In practice, bandpass filters, although functional in optical multiplexing, are a bulky WDM format. Several combinations of filters must be used depending on the number of channels multiplexed. Filter architectures could not be used for LSI board applications; however, some combination of WDM using a combination of two time division multiplexed (TDM) channels, where two channels are on different wavelengths, may be possible. It is not possible to wavelength multiplex a great number of channels using bandpass filters due to signal degradation encountered when passing through several filters. This is caused by the limited transmission ratio achievable with an interference filter. This ratio is 85% at best, although the maximum rejection of other wavelengths, determined by the reflection ratio, can be 100%.

2.7.4 TDM Schemes

TDM has important applications for LSI interconnects. When multiplexing a large number of signal channels, it is necessary to TDM to reduce interconnect complexity. Typical requirements are such that as many as 64 channels, each operating at 25 MHz, need to be transmitted. One possibility is to TDM the 16 bit address, data, and control words into four channels and then those four channels into one channel. This output channel would then be modulated at 1.6 GHz, which is entirely within the modulation rates attainable with laser diodes. Several commercially available GaAs universal shift regulators would meet the needs of this type of TDM.

2.8 Electro-Optic and Acousto-Optic Modulators

Electro-optic and acousto-optic modulators are the most highly developed types of modulators commercially available for digital communications. The drawback for these devices is that they operate in general on a power to bandwidth ratio of approximately 1 mW/MHz. The bandwidth of acousto-optic modulators is usually a small fraction of the acoustic frequency. This is highly prohibitive for data communications in the gigahertz band. The power requirements of these devices can also be prohibitive [82]. They are additionally quite bulky, although developments by some researchers to a rate above 1 Gb/s with order of magnitudes less drive power are available.

Electro-optic modulator response times are much more consistent with LSI data requirements. The electro-optic effect in certain materials is in the picosecond range and therefore does not limit the modulation bandwidth. The only limiting actors are the propagation times in the electro-optic crystals and the drive circuitry involved. For a modulator or reasonable drive voltage requirements, a bandwidth of approximately 2 GHz can be achieved at a wavelength of 0.633 μm in $LiNbO_3$ [83]. This type of electro-optic modulator also compares favorably in its dimensionality. Thus, electronically addressed electro-optic modulators are very useful for modulation of continuous sources within LSI interconnect systems.

2.9 Assessment of Interconnect System Architectures: Optical Networking Architectures

Network concepts for interconnecting a large number of computing elements are numerous, including rings, banyan, Clos, Omega, Benes', trees, spanning bus hypercubes, de Broijn, crossbar, gather scatter, cylinder, and other permutation networks. These networks must be compared to arrive at the most efficient network for interconnecting, given the ranges of data devices for a target application. "Efficient" in this context means not only the minimization of circuit complexity, delay, conflict at a node, synchronization, and coding, but also the ease of implementation in terms of optical interconnect technology.

There are several input issues that must be considered with high-speed optical interconnects. Many of these networks operate using burst-mode transmission. At very high-speeds, low levels of synchronization error or jitter are much more difficult to achieve if burst-mode transmission is used rather than synchronous operation. Jitter may be reduced by buffering, but at the cost of added network delay; therefore, synchronous operation is preferred to burst-mode transmission. Appropriate encoding schemes must be used for optical interconnects. Binary, balanced, and nonalphabetic codes are preferred.

2.9.1 Direct Relay Interconnects

The most direct method of interconnecting ICs from board to board is to interconnect the various data elements using LED arrays butt coupled to arrays or buses of fiber. This architecture can become inadequate, however, with large fan outs due to power requirements. Another direct-relay interconnect is the fiber star coupler. In this approach, fanout and selective interconnection is achieved from N ports to N other ports. This is done by twisting jacketed fibers while heating and pulling the fiber. When completed, this type of interconnect is an example of excellent passive interconnect, although the system gain requirements for the end to end response are prohibitive for LSI type fanouts utilizing laser diodes.

2.9.2 Fiber-Optic Data Busses

Fiber-optic data buses interconnect a number of spatially separated data points to permit any one point to communicate with any other point. They consist of a transmission medium, a mechanism for control of transmission over the medium, and an interface at each data location to provide a means for accessing the data bus. Data bus networks have several advantages over other network topologies such as point-to-point link. Most important, a data bus can make possible the interconnection of a set of points when the number of points is so large that interconnection through individual point-to-point links becomes impractical. Moreover, a data bus topology can provide considerable configurational flexibility; for example, points can be added to the network or moved to different locations without major revisions in the layout. These data buses are constructed using optical fiber and are similar to the star coupler architecture.

In particular, various data bus configurations can be constructed using optical fibers and biconical tapered structures connected by fusion splicing. The maximum number of terminals that can be interconnected for a given available optical budget between a transmitter and a receiver is dependent on source power. The connection between points is passive; active components such as light sources and detectors are located only at the fiber interfaces, so the failure of an active component at a fiber interface does not prevent the other points from using the data bus.

2.10 Interconnect Risk Assessments

Interconnect components or combinations of components present risks dependent on the ultimate data rate and bit error rate (BER) required in each application. The risks involved in implementing any technology in its infancy are always great, although certain interconnect methodologies and hardware carry greater risk than others due to implementation difficulties with respect to alignment tolerances, power consumption, BER, bandwidth, and other parameter inadequacies with respect to specific application architectures. This section endeavors to point out the architectures and hardware believed to best meet the present and future needs of optical interconnects.

Figure 2.32 illustrates hardware risk assessments and their probability of being utilized successfully in an LSI application optical interconnect. Figure 2.33 illustrates architecture risk assessments. Several things should

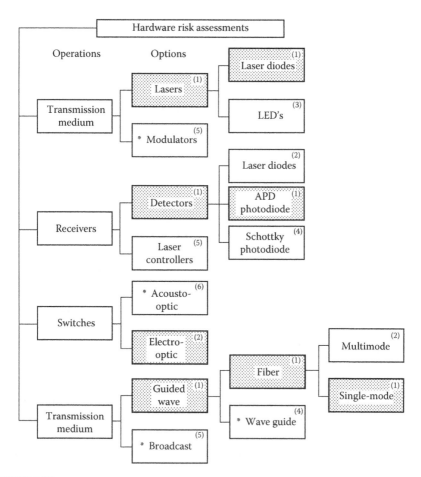

FIGURE 2.32
Hardware risk assessments.

be noted from the two figures. Many of the options for LSI interconnects are underdeveloped or have not been applied satisfactorily in application. These are the options marked with the asterisks in both figures. The ratings that have been assigned to all of the hardware and architectures are based on the probabilities of solution to implementation of optical data interconnects for intraboard and cabinet interconnections.

The development of key technologies has made LSI insertion into systems feasible using individual commercial components for optical data interconnects. Input and output terminals are dimensionally comparable to present day bonding pads. Power dissipation for optical elements can be made from a small fraction of total chip power dissipation and all high-speed data input and output can be performed by optical means. The conclusion is that optical interconnects are possible and sensible for intraboard and cabinet-level

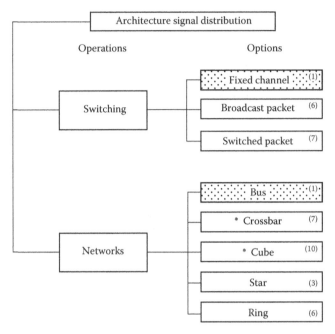

FIGURE 2.33
Architecture risks.

applications. There is no single optical interconnect approach suitable for every application, but rather each application will require separate consideration for the design of optical data links.

2.11 Electro-Optic System Applications

2.11.1 Special Application Lighting and Laser Illumination

An interesting test application for lasers is producing sufficient illumination to acquire high-speed video (frame rates up to a million frames per second) and to image through high-luminosity events such as fireballs. Previous laser illumination systems [84–87] required complex pulse timing, extensive cooling, large-scale laser systems (frequency-doubled flash-pumped Nd:YAG, Cu-vapor, Q-switched ruby), making them difficult to implement for illumination in high-speed videography. Requirements to illuminate through the self-luminosity of explosive events motivated the development of high-brightness imaging techniques obviating the limitations of previous attempts. Vertical cavity surface-emitting laser (VCSEL) arrays and a lens system were utilized with temporal and spectral filtering to effectively remove self-luminosity and fireball from the image, providing excellent background discrimination in a variety of range test scenarios.

Use of laser illumination in conjunction with narrow bandpass filtering and electro-optic shuttered temporal filtering provides background discrimination for high-resolution, high-speed digital image processing. In addition, the characteristics of the VCSEL structures now available represent viable lighting characteristics for other illumination events currently using flash-bulbs and argon discharge lamps. Timing sequences for shuttering, modulation, camera synchronization, exposure time, and digital image frame rate are integrated with a high degree of accuracy, permitting high resolution, blur- and distortion-free, quantitative video realization of high-speed events and associated data processing. Illumination from a VCSEL array is virtually speckle free. Measurements show that the illuminated area has speckle of <1%. Therefore, VCSEL illumination is very desirable for many applications where uniformity of illumination is very important.

Laser illumination can be applied in a broad range of applications. In some it may be necessary to pulse the laser for gated imaging. This technique can be used for 3-D imaging and also for improving the signal-to-noise ratio at a desired depth. In gated imaging, a synchronized triggering signal is sent to both the camera and the laser. Short pulses are then emitted from the laser and the reflected light from the scene is accepted at desired time delays. Other applications do not require gating.

Areas of interest may include

- Shaped charge detonations to further understand the properties of jet formation and particulation
- Explosively formed projectile detonations to quantify launch and flight performance characteristics
- Detonations of small caliber grenades and explosive projectiles to verify fuse function times and fragmentation patterns
- Performance and behavior of various projectiles and explosive threats against passive, reactive, and active target systems
- Human effects studies including body armor, helmets, and footwear
- Behind armor debris studies of large caliber ammunition against various armor materials
- Small caliber projectile firings to study launch, free flight, and target impact results
- Professional sporting events

Two key technologies are routinely utilized to achieve adequate scene illumination for high-speed video capture: flashbulbs and discharge tubes. Output curves for the frequently used PF-200 and PF-300 flashbulbs are shown in Figure 2.34. The peak output of 6.0×10^6 lm translates to 477,465 cd.

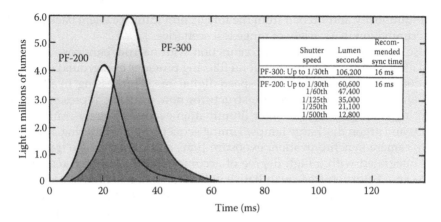

FIGURE 2.34
Flashbulb characteristics.

Bulbs are flashed, for instance, at 28 ms prior to the key portion of a video stream, so that for a 5- to 10-ms window, the target scene is illuminated at or above 90% of the peak flash output. The timing of the peak flash output and the character of the flash event were confirmed with .cine files obtained with a Vision Research Phantom V7.3 high-speed digital imaging system.

An example of the next-brightest light source, arc discharge tubes, currently identified and used for high-speed imaging, is the Luminys Corporation 6.5-K Blast high-intensity light. This source produces 25,863 cd, but for a significantly longer duration, depending on the charging characteristics. A typical illumination event for the 6.5-K Luminys light is shown in Figure 2.35.

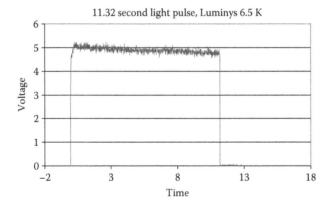

FIGURE 2.35
Time history of Luminys Corporation's 6.5-K light system.

2.11.2 Vertical Cavity Surface Emitting Lasers for Illumination

VCSELs were historically limited to low-power applications; but through several Defense Advanced Research Projects Agency (DARPA) sponsored programs [88], the output power of VCSELs dramatically improved. Princeton Optronics (PO) Corporation developed the world's highest power single devices and 2-D arrays [89]. They have successfully demonstrated single devices with >3 W continuous wave (CW) output power and large 2-D arrays with >230 W CW output power. These arrays comprise thousands of small, low-power, single-mode devices. The output of this array can then be focused into a very small, low-diverging spot using a microlens array focusing-lens system. VCSELs have several major advantages in illumination applications [89]:

- A circular output beam.
- Much lower wavelength dependence to temperature (5× less than for edge emitters). Since the VCSEL resonant cavity is defined by a wavelength-thick cavity sandwiched between two distributed Bragg reflectors (DBRs), devices emit in a single longitudinal mode and the emission wavelength is inherently stable (<0.07 nm/K), without the need for additional wavelength stabilization schemes or external optics, as is the case for edge emitters.
- A much higher reliability than edge emitters since VCSELs are not subject to catastrophic optical damage (COD) failures.
- High-temperature operation. VCSELs and VCSEL arrays can be operated at temperatures up to 80°C ambient without chillers.
- High power from the arrays (demonstrated CW power density of 1200 W/cm^2 from the arrays).
- Emission wavelength is very uniform across a 5 mm × 5 mm VCSEL array, resulting in spectral widths of ~0.7 to 0.8-nm full width at half maximum (FWHM).
- VCSELs can be easily processed into large 2-D arrays (do not need stacking) to scale up the power. The 2-D array configuration provides for more efficient heat-sinking and better pump power density, as the devices can be very closely packed.
- Low thermal resistance for CW operation. A thermal resistance of 0.15°C/W from the junction temperature to the microcooler coolant makes it lower by more than a factor of 2 compared with edge-emitter arrays.
- VCSEL arrays are designed for coupling match to COTS microlens arrays.
- PO makes single-mode devices at wavelengths of 976 and 1064 nm, as well as custom wavelengths between 808 and 1064 nm.

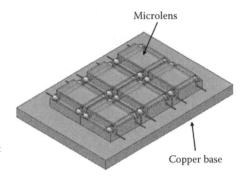

FIGURE 2.36
Six-VCSEL array configuration (current configurations contain nine).

As a starting point, we consider the optical output needed to achieve a luminous intensity comparable to that of the arc discharge lamp (25,000 lm). A single-VCSEL array will deliver 70 W at 976 nm. With 1 W = 680 lm at 550 nm, and a correction factor of 7 (assuming a camera QE of 35% at 550 nm and 5% at 976 nm), the equivalent single-VCSEL output at 976 nm is about 6800 lm. A 9-VCSEL device will deliver about 61,200 lm of focused light onto its target. It can be expected that the VCSEL will improve lighting by more than one f-stop, and the source can be used at extended distances.

The arrays are connected serially and mounted on a copper base (Figure 2.36). For 150-ms operation, the heat released is about 95 J. The specific heat of copper is 0.385 J/g°K; therefore, for 100 g of copper submount, the temperature increase is only 2.5°C, which allows passive cooling.

Nebolsine et al. [85] developed estimates for shaped charge jets and radiometric characterization with video cameras. The approach was to determine illumination levels for a laser illuminator needed to overcome the jet luminosity and provide sufficient exposure for imaging. This general concept was used to evaluate candidate laser technologies for suitability in system design. The VCSEL approach provides flexibility, in that it can be utilized in a range of timing and illumination test scenarios. The pulse length can be essentially CW, with minimal cooling requirements; while the illumination brightness can be increased through the addition of more arrays in the package configuration.

2.11.3 Spectral Matching Considerations

The monochrome spectral response curve for the Vision Research Phantom V7.3 high-speed video camera used in the author's laboratory is shown as the dark curve in Figure 2.37. While light sources such as the Luminys 6.5-K lamp exhibit broadband spectral content (Figure 2.38), similar to the

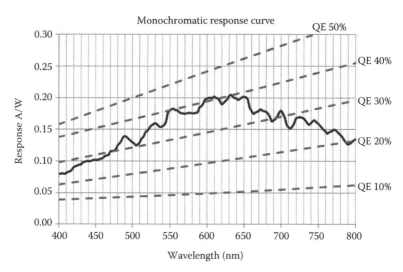

FIGURE 2.37
Spectral sensitivity of Phantom v7.3 high-speed digital imaging system.

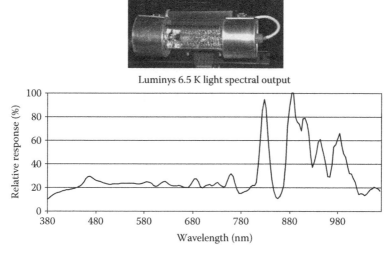

FIGURE 2.38
Normalized spectral output of Luminys Corp. 6.5-K light.

flashbulbs, VCSEL sources produce a 976-nm center wavelength, with a FWHM spectral width of 0.8 nm. For this reason, the response of the V7.3 at 976 nm is an important parameter for our system design. As mentioned previously, the ratio of the quantum efficiency at 550 nm to the efficiency at 976 nm is about 7.

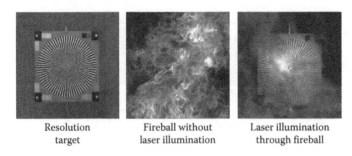

| Resolution | Fireball without | Laser illumination |
| target | laser illumination | through fireball |

FIGURE 2.39
Laboratory demonstration of single-VCSEL illuminator.

2.11.4 High-Brightness Imaging

The system developed is intended for a wide variety of illumination scenarios. The VCSEL can be pulsed and operated uncooled for the duration of short events (e.g., 150–200 ms), as was the case with previous work in laser illuminators, where the light source determined the temporal sample size for the event. For longer event applications, it can be operated CW and in a cooled configuration. In CW operation, the temporal characteristics of high-speed capture are determined by the camera.

For imaging applications involving the liner collapse, jet formation, and progression of a detonated shaped charge, some camera timing capabilities may not be fast enough to fully characterize the event. Typical jet velocities approach 10–12 km/s [85, 92]. In these cases, one might consider, for instance, the Imacon 200 ultra high-speed camera, with frame rates up to 200 million frames per second, or 5-ns samples.

Use of VCSELs for laser illumination applications shows great promise as a versatile, readily implemented approach to a host of current challenges for high-speed imaging. A technology demonstration phase utilizing a single-VCSEL array has been performed in the author's laboratory. A view from the demonstration is shown in Figure 2.39. This illustrates the ability to see through a fireball using laser illumination and narrow bandpass filtering. The utility of the VCSEL illumination is readily apparent. A fielded brass board 9-VCSEL model, shown below in Figure 2.40, is now in testing in live fire applications.

2.12 Vertical Cavity Surface Emitting Laser Technology

Princeton Optronics (PO), in Trenton, New Jersey, is the leader in commercial VCSEL technology. The following information (Section 2.12) is courtesy of PO and Dr. Chuni Ghosh.

FIGURE 2.40
Nine-VCSEL laser illuminator.

2.12.1 Introduction

VCSELs are a relatively recent type of semiconductor laser. These were first invented in the mid-1980s. Very soon, they gained the reputation as a superior technology for short reach applications such as fiber channel, Ethernet, and intrasystems links. Within the first 2 years of commercial availability (1996), VCSELs became the technology of choice for short range datacom and local area networks, effectively displacing edge-emitter lasers. Mentzer, Naghski, and a team of IBM researchers developed the first VCSEL-based parallel 12-channel fiber-optic data link [93]. This success was mainly due to the VCSEL's lower manufacturing costs and higher reliability compared to edge emitters.

PO developed the key technologies resulting in the world's highest power single-VCSEL devices and 2-D arrays, demonstrating single devices with >5 W CW output power and large 2-D arrays with >230 W CW output power. They made single-mode devices of 1 W output power and single-mode arrays with power of >100 W, which are coupled to 100u, 0.22NA fiber. The highest wall plug efficiency of these devices and arrays is 56%. Arrays delivered 1 kW/cm^2 in CW operation and 4.2 kW/cm^2 in QCW operation. PO participated in the DARPA-SHEDS program, whose main objective was to improve laser diode power conversion efficiency.

2.12.2 VCSEL Structure

Semiconductor lasers consist of layers of semiconductor material grown on top of each other on a substrate (the "epi"). For VCSELs and edge emitters, this growth is typically done in an MBE or MOCVD growth reactor. The grown wafer is then processed accordingly to produce individual devices. Figure 2.41 summarizes the differences between VCSEL and edge-emitter processing.

In a VCSEL, the active layer is sandwiched between two highly reflective mirrors (dubbed DBRs) made up of several quarter-wavelength-thick layers of semiconductors of alternating high and low refractive index. The reflectivity of these mirrors is typically in the range of 99.5%–99.9%. As a result, the light oscillates *perpendicular* to the layers and escapes through the top (or bottom) of the device. Current and/or optical confinement is typically achieved through either selective-oxidation of an aluminum-rich layer, ion-implantation, or even both for certain applications. The VCSELs can be designed for "top-emission" (at the epi–air interface) or "bottom-emission" (through the transparent substrate), in cases where "junction-down" soldering is required for more efficient heat sinking, for example. Figure 2.42 illustrates different common types of VCSEL structures.

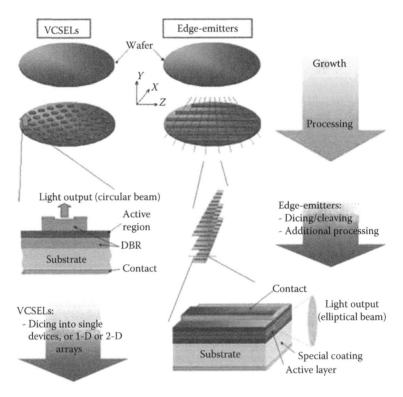

FIGURE 2.41
Comparison of the growth/processing flow of VCSEL and edge-emitter semiconductor lasers. (Courtesy of Princeton Optronics [PO] and Dr. Chuni Ghosh. With permission.)

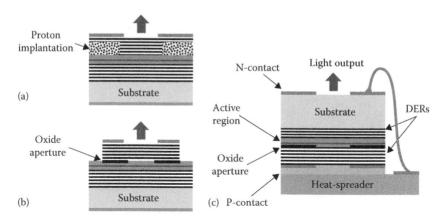

FIGURE 2.42

Three common types of VCSEL structures: (a) a top-emitting structure with proton implantation to confine the current, (b) a selectively oxidized top-emitting structure to confine the optical modes and/or the current, and (c) a mounted bottom-emitting selectively oxidized structure. (Courtesy of Princeton Optronics [PO] and Dr. Chuni Ghosh. With permission.)

In contrast, edge emitters are made up of cleaved bars diced from the wafers. Because of the high index of refraction contrast between air and the semiconductor material, the two cleaved facets act as mirrors. Hence, in the case of an edge emitter, the light oscillates *parallel* to the layers and escapes sideways. This simple structural difference between the VCSEL and the edge emitter has important implications.

Since VCSELs are grown, processed, and tested while still in the wafer form, there is significant economy of scale resulting from the ability to conduct parallel device processing, whereby equipment utilization and yields are maximized, and setup times and labor content are minimized. In the case of a VCSEL (see Figure 2.41), the mirrors and active region are sequentially stacked along the Y-axis during epitaxial growth. The VCSEL wafer then goes through etching and metallization steps to form the electrical contacts. At this point, the wafer goes to test where individual laser devices are characterized on a pass-fail basis. Finally, the wafer is diced and the lasers are binned for either higher level assembly (typically >95%) or scrap (typically <5%). Figure 2.43a shows a single high-power VCSEL device (>2 W output power) packaged on a high-thermal conductivity submount. Figure 2.43b shows the $L–I$ characteristics of a 5 W VCSEL device.

In a simple Fabry–Pérot edge emitter, the growth process also occurs along the Y-axis, but only to create the active region, as mirror coatings are later applied along the Z-axis. After epitaxial growth, the wafer goes through the metallization step and is subsequently cleaved along the X-axis, forming a series of wafer strips. The wafer strips are then stacked and mounted into a coating fixture. The Z-axis edges of the wafer strips are then coated to form

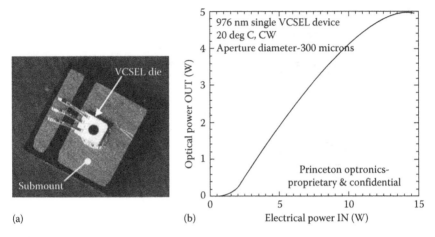

(a)   (b)   Electrical power IN (W)

FIGURE 2.43
(a) 3 Packaged high-power VCSEL device (>2W). The submount is 2mm × 2mm. (b) Shows the L-I characteristics of a 5W VCSEL device. The device aperture is 300u. (Courtesy of Princeton Optronics [PO] and Dr. Chuni Ghosh. With permission.)

the device mirrors. This coating is a critical processing step for edge emitters, as any coating imperfection will result in early and catastrophic failure of the devices due to COD. After this coating step, the wafer strips are diced to form discrete laser chips, which are then mounted onto carriers. Finally, the laser devices go into test.

It is also important to understand that VCSELs consume less material: in the case of a 3″ wafer, a laser manufacturer can build about 15,000 VCSEL devices or approximately 4,000 edge emitters of similar power levels. In addition to these advantages, VCSELs demonstrate excellent dynamic performance such as low threshold current (a few microamperes), low-noise operation, and high-speed digital modulation (10 Gb/s). Furthermore, although VCSELs have been typically utilized in low-power applications—a few milliwatts at most—they have the inherent potential of producing very high powers by processing large 2-D arrays, as discussed in the section describing high-brightness illumination. In contrast, edge emitters cannot be processed in 2-D arrays.

2.12.3 VCSEL Advantages

The many additional advantages offered by the VCSEL technology can be summarized in the following points:

1. Wavelength stability: The lasing wavelength in a VCSEL is very stable, since it is fixed by the short (1- to 1.5-wavelength-thick) Fabry–Pérot cavity. Contrary to edge emitters, VCSELs can only operate in a single longitudinal mode.

2. Wavelength uniformity and spectral width: Growth technology has improved such that VCSEL 3" wafers are produced with <2 nm standard deviation for the cavity wavelength. This allows the fabrication of VCSEL 2-D arrays with little wavelength variation between the elements of the array (<1 nm FWHM spectral width). By contrast, edge-emitter bar-stacks suffer from significant wavelength variations from bar to bar since there is no intrinsic mechanism to stabilize the wavelength, resulting in a wide spectral width (3–5 nm FWHM).

3. Temperature sensitivity of wavelength: The emission wavelength in VCSELs is approximately five times less sensitive to temperature variations than in edge emitters. The reason is that in VCSELs, the lasing wavelength is defined by the optical thickness of the single-longitudinal mode cavity and that the temperature dependence of this optical thickness is minimal (the refractive index and physical thickness of the cavity have a weak dependence on temperature). On the other hand, the lasing wavelength in edge emitters is defined by the peak-gain wavelength, which has a much stronger dependence on temperature. As a consequence, the spectral line-width for high-power arrays (where heating and temperature gradients can be significant) is much narrower in VCSEL arrays than in edge-emitter arrays (bar-stacks). Also, over a 20°C change in temperature, the emission wavelength in a VCSEL will vary by <1.4 nm (compared to ~7 nm for edge emitters).

4. High-temperature operation (Chillerless operation for pumps): VCSEL devices can be operated without refrigeration because they can be operated at temperatures to 80°C, the cooling system becomes very small, rugged, and portable with this approach.

5. Higher power per unit area: Edge emitters deliver a maximum of about 500 W/cm² because of gaps between bar to bar that must be maintained for coolant flow, while VCSELs are delivering ~1200 W/cm² now and will deliver 2–4 kW/cm² in the near future.

6. Beam quality: VCSELs emit a circular beam. Through proper cavity design, VCSELs can also emit in a single transverse mode (circular Gaussian). This simple beam structure greatly reduces the complexity and cost of coupling/beam-shaping optics (compared to edge emitters) and increases the coupling efficiency to the fiber or pumped medium. This has been a key selling point for the VCSEL technology in low-power markets.

7. Reliability: Because VCSELs are not subject to COD, their reliability is much higher than for edge emitters. Typical FIT values (failures in one billion device-hours) for VCSELs are <10.

8. Manufacturability and yield: Manufacturability of VCSELs has been a key selling point for this technology. Because of complex manufacturing processes and reliability issues related to COD, edge emitters have a low yield (edge-emitter 980-nm pump chip manufacturers typically only get ~500 chips out of a 2" wafer). On the other hand, yields for VCSELs exceed 90% (corresponds to ~5000 high-power chips from a 2" wafer). In fact, because of its planar attributes, VCSEL manufacturing is identical to standard IC Silicon processing.

9. Scalability: For high-power applications, a key advantage of VCSELs is that they can be directly processed into monolithic 2-D arrays, whereas this is not possible for edge emitters (only 1-D monolithic arrays are possible). In addition, a complex and thermally inefficient mounting scheme is required to mount edge-emitter bars in stacks.

10. Packaging and heat-sinking: Mounting of large high-power VCSEL 2-D arrays in a "junction-down" configuration is straightforward (similar to microprocessor packaging), making the heat-removal process very efficient, as the heat must traverse only a few microns of AlGaAs material. Record thermal impedances of <0.16 K/W have been demonstrated for 5 mm × 5 mm 2-D VCSEL arrays.

11. Cost: With the simple processing and heat-sinking technology, it becomes much easier to package 2-D VCSEL arrays than an equivalent edge-emitter bar-stack. The established silicon industry heat-sinking technology can be used for heat removal for very high-power arrays. This will significantly reduce the cost of the high-power module. Currently, cost of the laser bars is the dominant cost for the DPSS lasers.

12. High-wavelength stability and low-temperature dependence: Since the VCSEL resonant cavity is defined by a wavelength-thick cavity sandwiched between two DBRs, devices emit in a single longitudinal mode and the emission wavelength is inherently stable (<0.07 nm/K), without the need for additional wavelength stabilization schemes or external optics, as is the case for edge emitters. Furthermore, thanks to advances in growth and packaging technologies, the emission wavelength is very uniform across a 5 mm × 5 mm VCSEL array, resulting in spectral widths of 0.7–0.8 nm (FWHM). (see Figures 2.44 and 2.45.)

Wavelength stability and narrow spectral width are very significant advantages in pumping applications, for example, where the medium has a narrow absorption band.

Unlike edge emitters, VCSELs emit in a circularly symmetric beam with low divergence without the need for additional optics. This has

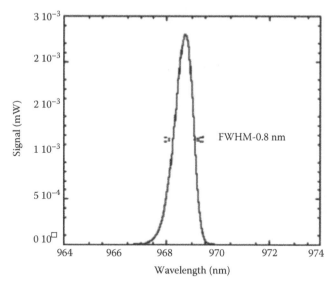

FIGURE 2.44
Emission spectrum of a 5 mm × 5 mm VCSEL array at 100 W output power (120 A). (Courtesy of Princeton Optronics [PO] and Dr. Chuni Ghosh. With permission.)

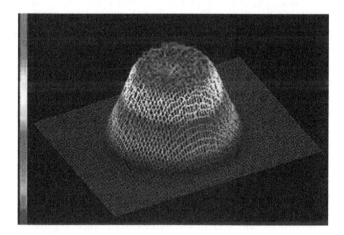

FIGURE 2.45
Far-field beam profile of a 5 mm × 5 mm VCSEL array at 100 W output power (120 A). (Courtesy of Princeton Optronics [PO] and Dr. Chuni Ghosh. With permission.)

been a tremendous advantage for low-power VCSELs in the telecom and datacom markets because of their ability to directly couple to fibers ("butt-coupling") with high coupling efficiency. PO high-power VCSEL arrays emit in a quasi-top-hat beam profile, making these devices ideal for direct pumping ("butt-pumping") of solid-state lasers.

13. Feedback insensitivity: In VCSELs, the as-grown output coupler reflectivity is very high (typically >99.5%) compared to edge emitters (typically <5%). This makes VCSELs extremely insensitive to optical feedback effects, thus eliminating the need for expensive isolators or filters in some applications.

14. Low thermal impedance and ease of packaging: PO developed advanced packaging technologies, enabling efficient and reliable die-attach of large 2-D VCSEL arrays on high-thermal-conductivity submounts. The resulting submodule layout allows for straightforward packaging on a heat exchanger.

For high-power devices packaged on microcoolers, PO demonstrated modules with thermal impedances as low as 0.15 K/W (between the chiller and the chip active layer).

2.12.4 High-Power CW and QCW VCSEL Arrays

PO designs and manufactures advanced high-power CW and QCW diode lasers for the industrial, medical, and defense markets. Unlike edge emitters, the light emits perpendicular to wafer surface for VCSELs. It is therefore a straightforward to process 2-D arrays of small VCSEL devices driven in parallel to obtain higher output powers. The advantage of 2-D arrays is the simple silicon IC chip-like configuration. Many of the silicon IC packaging and cooling technologies can be applied to VCSEL arrays.

PO took the VCSEL technology to very high power levels by developing very large (5 mm × 5 mm) 2-D VCSEL arrays packaged on high-thermal-conductivity submounts. (see Figure 2.46). These arrays are composed of thousands of low-power single devices driven in parallel. Using this approach, record CW output powers in excess of 230 W from a 0.22 cm^2 emission area (>1 kW/cm^2) have been demonstrated, without sacrificing wall-plug efficiency.

In addition to CW VCSEL arrays, PO has developed very high power density VCSEL arrays for quasi-CW (QCW) operation. QCW powers in excess of 925 W have been demonstrated from very small arrays (5 × 5 mm chip size), resulting in record power densities >4.2 kW/cm^2 (see Figure 2.47). These small arrays can easily be connected in series to form larger arrays with high output powers. These arrays are ideal for applications requiring very compact high-power laser sources.

Because VCSELs can operate reliably at temperatures up to 80°C, they do not necessarily require refrigeration. Additionally, since the wavelength change with temperature is small, the cooling system design can be considerably simplified. The cooling system thus becomes very small, rugged, and portable with this approach. We have been operating the VCSELs and VCSEL

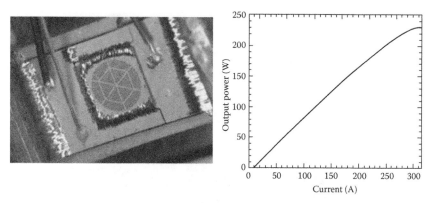

FIGURE 2.46
Picture of high-power 5 mm × 5 mm 2-D VCSEL array mounted on a micro-cooler and measure CW output power and voltage at a constant heat-sink temperature. Roll-over power is >230 W. (Courtesy of Princeton Optronics [PO] and Dr. Chuni Ghosh. With permission.)

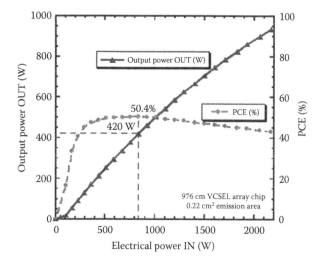

FIGURE 2.47
Power versus current for a small VCSEL 2-D array under different QCW regimes. These arrays exhibit power densities of >4.2 kW/cm². (Courtesy of Princeton Optronics [PO] and Dr. Chuni Ghosh. With permission.)

arrays with water pump and a radiator cooling like that of a car engine. Figure 2.48 shows such a setup in which a radiator and water pump is used to cool a 120 W array of VCSELs. The result of the cooling arrangement is compared with a chiller cooling and shown in Figure 2.43.

Figure 2.49a shows the performance of a 120 W VCSEL array with fan-radiator cooling with water temperature at 45°C versus cooling with a

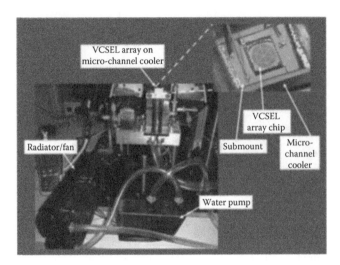

FIGURE 2.48
Shows the setup without chiller using a radiator and a water pump in an arrangement like in a car engine.

chiller with 16°C water temperature. The green curves show the efficiency (CE) in the two cases, which is almost similar. The red curves show the power output from the array in the two cases. The power output decreases somewhat at higher power, but at power levels below 80 W, there is very little change. Figure 2.49b shows the L–I curves for a QCW array at 808 nm at different temperatures. The top curve is at 20°C and the bottom one is at 85°C.

2.12.5 VCSEL Reliability

In terms of reliability, VCSELs have an inherent advantage over edge emitters because they are not subject to COD. Indeed, the problem of sensitivity to surface conditions for edge emitters is not present in VCSELs because the gain region is embedded in the epistructure and does not interact with the emission surface.

Over the years, several reliability studies for VCSELs have yielded FIT rates (number of failures in one billion device-hours) on the order of 1 or 2, whereas FIT rates for the highest telecom-grade edge emitters is on the order of 500. The failure rate for industry-grade high-power edge-emitter bars or stacks is even worse (>1000).

PO has accumulated millions of device-hours on VCSELs operating above 100°C. This reliability advantage will be very significant for laser systems, where the end-of-life and field failures are overwhelmingly dominated by pump failure. Moreover, VCSEL arrays can be operated at higher

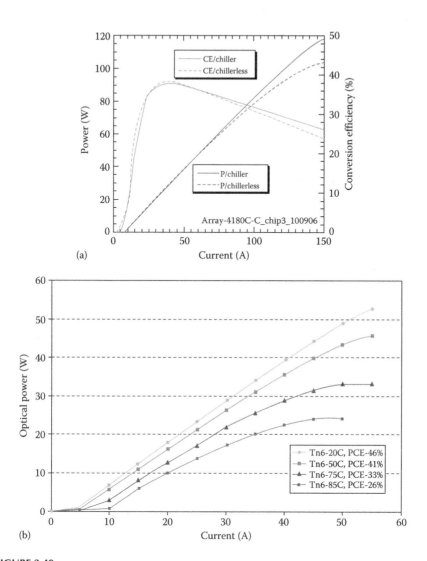

FIGURE 2.49
(a) Performance of a 120W VCSEL array with fan-radiator cooling. (b) *L–I* curves for a QCW array at 808nm at different temperatures. (Courtesy of Princeton Optronics [PO] and Dr. Chuni Ghosh. With permission.)

temperatures, resulting in lower power consumption of the overall laser system.

2.12.6 Single-Mode VCSEL Devices

PO makes single-mode devices with power levels of 5 mW for small-aperture devices (4μm) as well as higher power devices of power output of 150mW from 100 μm aperture devices in TO can packages and up to 1 W from a single device in a larger package. The devices have narrow linewidths of tens of kilohertz and SMSR of >30 dB. The devices have high efficiency of >40%.

2.12.7 High-Speed VCSEL Devices

High-speed VCSEL devices are in development. Current devices work to 5 GHz speed. Figure 2.50 shows the performance of a 4 μm aperture device performing at >5 GHz speed. A power level of >5 mW is achieved at these speeds.

2.12.8 High-Brightness Arrays of Single-Mode Devices

VCSELs are also a new technology for pumping of solid state lasers, including fiber lasers. Because of their circular beam and excellent optical characteristics, it is possible to make high-brightness pumps using an array of closely spaced single-mode VCSEL devices. The arrays are temperature stable and operate at high temperatures without chillers. They are lower cost as they do not require expensive operations like cleaving of the wafers. These fiber coupled arrays can be combined for higher output power using fiber combiners and coupled with double core fiber to develop small, high-performance fiber lasers working at elevated temperatures without chillers.

For high-brightness devices and arrays, PO makes self-lasing VCSELs or extended or external cavity VCSELs and couples such arrays into fibers using microlenses as shown in Figure 2.51.

2.12.9 Blue, Green, and UV VCSELs

PO developed very high quality blue lasers from VCSELs by frequency doubling the VCSEL radiation with a nonlinear material. The blue laser output at

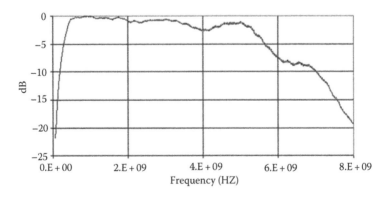

FIGURE 2.50
A 4 u aperture single-mode device shows high-speed performance through >5 GHz. The device power is >5 mW.

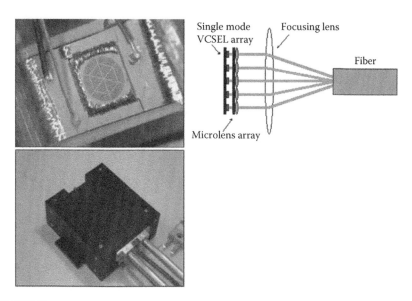

FIGURE 2.51
This figure shows a 100 W CW single-mode VCSEL array mounted on a microcooler (top left). On top right, it shows the single-mode array coupling scheme to couple the array with the fiber. At bottom left, it shows the package that has a dimension of (2 × 1.5 × 0.5″), which is water cooled. (Courtesy of Princeton Optronics [PO] and Dr. Chuni Ghosh. With permission.)

480 nm is single-mode and highly monochromatic with a beam divergence (half angle) of 8 mR. The VCSEL devices and arrays are capable of delivering very high power in a 2-D array and hence frequency doubled arrays are able to deliver very high levels of power. PO developed 6 W of peak power from a single-VCSEL device that was frequency doubled and is working toward a 10 mJ pulse, 1 kHz rep rate blue laser in a small form factor. Figure 2.52 shows the schematic of the approach and Figure 2.53 shows the experimental setup for the blue laser using VCSELs.

For UV lasers, an external frequency doubler material BGO doubles the frequency of the laser. PO is developing several millijoules of UV energy per pulse from these devices with kilohertz repetition rate. Figure 2.54 shows the schematics of the approach obtaining 1 mJ pulse energy, with a 100 Hz repetition rate.

2.12.10 Narrow Divergence Arrays

VCSEL arrays with microlens collimation become very narrow divergence light sources. A single-mode array (Figure 2.55) shows the collimation of the beams from individual VCSELs. Using a microlens aligned with the VCSEL array and held in position by laser-welding the holding frames produces divergence of 60 mrad (full angle) for self-lasing arrays and 8 mrad (full angle) for external cavity arrays. Figure 2.56 shows schematically the

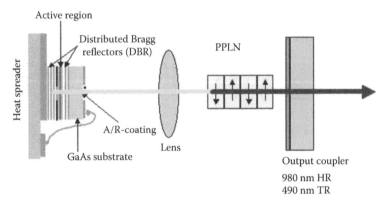

FIGURE 2.52
This figure shows the schematic of blue laser generation. An external cavity approach is used in which the PPLN is put inside the cavity. We have developed green laser using this approach as well. A peak power of 6 W from a single VCSEL has been obtained with this approach. (Courtesy of Princeton Optronics [PO] and Dr. Chuni Ghosh. With permission.)

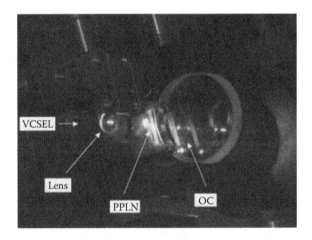

FIGURE 2.53
This figure shows the experimental setup for frequency doubling of VCSEL devices. PPLN material is used for frequency doubling. PO is working on frequency doubling of high-power arrays. A peak power of 6 W from a single device has been obtained. (Courtesy of Princeton Optronics [PO] and Dr. Chuni Ghosh. With permission.)

collimation architecture using microlenses. Using external lenses as shown in Figure 2.56, one can achieve very narrow divergence with VCSEL arrays. A divergence of 0.5 mrad has been achieved using an expander lens and a focusing lens.

As discussed previously, VCSELs provide excellent illumination sources for high-luminosity lighting requirements, such as high-speed videography. PO developed an area illuminator module at 808 nm that can be used

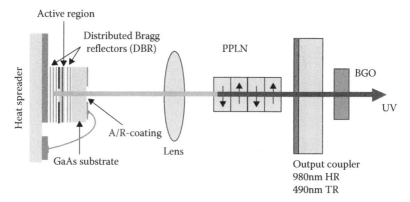

FIGURE 2.54
This figure shows the schematics of the frequency-doubling approach to UV wavelengths. We are working toward developing lasers with several millijoules per pulse with 1 kHz repetition rate. (Courtesy of Princeton Optronics [PO] and Dr. Chuni Ghosh. With permission.)

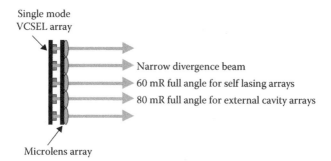

FIGURE 2.55
This figure shows the architecture of collimating the individual beams from the VCSELs in the array by means of a microlens array. (Courtesy of Princeton Optronics [PO] and Dr. Chuni Ghosh. With permission.)

to illuminate a large area. The module has a power output 400 W and can be used to illuminate an area of 1.5 km × 1.5 km. Figures 2.57 and 2.58 show the diagram of the module as well as the picture of it.

Similar VCSEL illuminators are used with silicon CCD or CMOS cameras for illumination for perimeter security, area illumination, border security etc. Figures 2.59 and 2.60 show diagrams for these illuminators that are small form factor, have high efficiency, and have low cost.

2.12.11 VCSEL-Based 1064 nm Low-Noise Laser

PO developed a novel approach for low-noise lasers with extremely low noise using VCSELs. For high-power operation, efficient heat removal is required

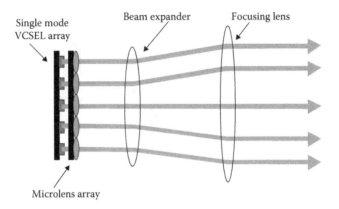

FIGURE 2.56
This figure shows the architecture of collimating the individual beams from the VCSELs and further reducing their divergence by using a beam expander and a focusing lens. A divergence of 0.5 mR can be achieved from the entire array with this approach. (Courtesy of Princeton Optronics [PO] and Dr. Chuni Ghosh. With permission.)

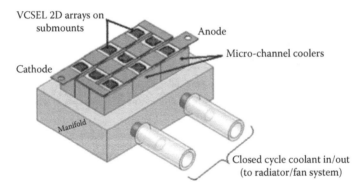

FIGURE 2.57
The diagram of the 400 W illuminator module. The dimensions of the module is 2.5 × 1.5 × 5.5″. (Courtesy of Princeton Optronics [PO] and Dr. Chuni Ghosh. With permission.)

and therefore a junction-down, bottom-emitting structure is preferred to improve current injection uniformity in the active region and to reduce the thermal impedance between the active region and the heat spreader. A schematic of the structure without the heat spreader is shown in Figure 2.61.

For current and optical confinement, the selective oxidation process is used to create an aperture near the active region to improve performance. A low-doped GaAs N-type substrate is used to minimize absorption of the output light while providing electrical conductivity for the substrate-side N-contact. The growth is performed in a MOCVD or MBE reactor and starts with an AlGaAs N-type partially reflecting DBR. The active region consists of InGaAs quantum wells designed for 1064 nm emission and

FIGURE 2.58
Picture of the module. (Courtesy of Princeton Optronics [PO] and Dr. Chuni Ghosh. With permission.)

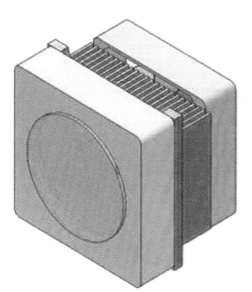

FIGURE 2.59
Diagram of a 3 and 8 W illuminator (808 nm). The dimensions of the illuminator is $2 \times 2 \times 2''$ and uses 15 W input power for 3 W output and 40 W for 8 W output. The beam divergence is 16°. (Courtesy of Princeton Optronics [PO] and Dr. Chuni Ghosh. With permission.)

strained-compensated using GaAsP barriers. The active region is followed by a high-reflecting P-type DBR. A high-aluminum content layer is placed near the first pair of the P-DBR to later form the oxide aperture. The placement and design of the aperture are critical to minimize optical losses and current spreading. Band-gap engineering (including modulation doping) is used to design low-resistivity DBRs with low-absorption losses.

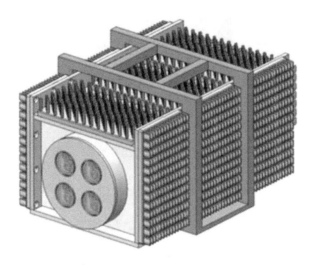

FIGURE 2.60

Diagram of a 40 W output illuminator. The input power for the illuminator is 200 W. The dimension of the illuminator is $7.5 \times 6.5 \times 12$ in. (Courtesy of Princeton Optronics [PO] and Dr. Chuni Ghosh. With permission.)

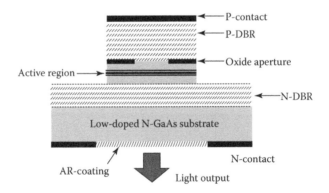

FIGURE 2.61

Schematic of the selectively oxidized, bottom-emitting 1064 nm VCSEL structure. (Courtesy of Princeton Optronics [PO] and Dr. Chuni Ghosh. With permission.)

2.12.12 Low-Noise Laser Cavity

The low-noise laser cavity is shown in Figure 2.62. The optical setup consists of a VCSEL device and an output coupler. A high-quality Etalon and a Brewster plate in the cavity control the single-wavelength operation and linear polarization, respectively. The optical isolator prevents optical feedback from outside of the cavity to maintain the single-wavelength operation. The beam is then coupled into a PM fiber with a focusing lens for the mode matching.

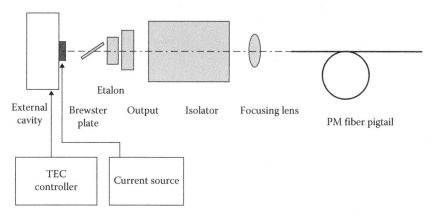

FIGURE 2.62
The low-noise laser cavity with an etalon and a Brewster plate can be seen. (Courtesy of Princeton Optronics [PO] and Dr. Chuni Ghosh. With permission.)

2.13 Derivation of the Linear Electro-Optic (Pockels) Effect

The linear electro-optic effect is the result of distortion of the crystal lattice caused by an applied electric field. The effect is manifested as an induced birefringence in the crystal that results in field dependent changes in the refractive index along various crystal axes, thus affecting the phase of the transmitted electric field. The electro-optic effect can be derived using the Index Ellipsoid (IE) notation. A complete derivation for 43 m crystals can be found in Namba and an example can be found in Yariv [94]. The procedure is to transform the standard IE to a coordinate system rotated to match the symmetry of the field of interest.

As an example, Figure 2.63 shows an exaggerated diagram of a Double-Y Mach Zehnder optical logic gate [95] (see also Chapter 3) on a GaAs wafer, indicating the relative orientation of the device and the crystal. The crystal axes are indicated using standard notation. It should be noted that due to the relationship between wafer orientation and the crystal axis of rotation, it is only practical to apply an external field along a single crystal axis.

The standard IE is written

$$IE = \frac{1}{n_0^2}(x^2 + y^2 + z^2) + 2r_{41}(E_x yz + E_y xz + E_x xy) \tag{2.7}$$

For this discussion, we assume the applied dc field is in the x direction only, along the <100> axis; therefore,

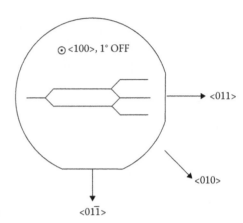

FIGURE 2.63
Wafer orientation of double-Y device.

$$E = E_{x'} \quad \text{and} \quad \text{IE} = \frac{1}{n_0^2}(x^2 + y^2 + z^2) + 2r_{41}(E_x yz) \tag{2.8}$$

We transform the standard IE to primed coordinates. By symmetry

$$x' = x'' \quad x = x' \quad x^2 = x'^2 \tag{2.9}$$

By inspection

$$y' = \frac{y}{\sqrt{2}} + \frac{z}{\sqrt{2}}; \quad y = \frac{y'}{\sqrt{2}} - \frac{z'}{\sqrt{2}} \tag{2.10}$$

$$z' = \frac{(-y)}{\sqrt{2}} + \frac{z}{\sqrt{2}}; \quad z = \frac{y'}{\sqrt{2}} + \frac{z'}{\sqrt{2}}$$

$$y^2 = \frac{y'^2}{2} + \frac{z'^2}{2} + y'z'$$

$$z^2 = \frac{y'^2}{2} + \frac{z'^2}{2} + y'z'$$

$$yz = \frac{y'^2}{2} - \frac{z'^2}{2}$$

Substituting into IE,

$$IE = x'^2\left(\frac{1}{n_0^2}\right) + y'^2\left(\frac{1}{n_0^2} + r_{41}E_x\right) + z'^2\left(\frac{1}{n_0^2} - r_{41}E_x\right) \qquad (2.11)$$

The standard primed IE′ is

$$IE' = \frac{x'^2}{n_x'^2} + \frac{y'^2}{n_y'^2} + \frac{z'^2}{n_z'^2} \qquad (2.12)$$

Equating IE = IE′ and assuming $1 \gg n_0^2 r_{41}E_x$ yields

$$n_s' = n_0 \qquad (2.13)$$

$$n_y' = n_0\left(1 - \frac{r_{41}n_0^2}{2}E_x\right)$$

$$y' = \frac{y}{\sqrt{2}} + \frac{z}{\sqrt{2}} \gg y = \frac{y'}{\sqrt{2}} - \frac{z'}{\sqrt{2}}$$

$$n_x' = n_0\left(1 + \frac{r_{41}}{2}n_0^2 E_x\right)$$

$$z' = \frac{-y}{\sqrt{2}} + \frac{z}{\sqrt{2}} = z = \frac{y'}{\sqrt{2}} + \frac{z}{\sqrt{2}}$$

$$y^2 = \frac{y'^2}{2} + \frac{z'^2}{2} - y'z'$$

$$z' = \frac{y'^2}{2} + \frac{z'^2}{2} + y'z'$$

$$yz = \frac{y'^2}{2} - \frac{z'^2}{2}$$

This index change is

$$\Delta n = n_i' - n_i' = n_0 \qquad (2.14)$$

or

$$\Delta n_x = 0$$

$$\Delta n_y' = -r_{41} \frac{n_0^3}{2} E_x$$

$$\Delta n_x' = +r_{41} \frac{n_0^3}{2} E_x$$

The results of this analysis show that one must consider both the waveguide alignment and the polarization of the optical field launched in the waveguide. This shows that an index change is encountered only by field components lying in the plane of the wafer. Furthermore, the index change depends on the crystal direction of the optical propagation.

The index change will vary from positive along the directions <011> or $<0\bar{1}1>$ to negative along the directions $<01\bar{1}>$ and $<0\bar{1}\bar{1}>$; along four of the crystal axes, <001>, $<00\bar{1}>$, <010>, and $<0\bar{1}0>$, the positive and negative changes cancel and there is no net index change.

To determine the ideal waveguide direction, two points must be considered. First, arbitrary polarization of the launched E-field will generally result in an induced elliptical polarization. Only by launching the field parallel to the wafer will the linear polarization be maintained. The second consideration is that the waveguiding of any optical field is accomplished by making the guiding region a higher index of refraction than the cladding; therefore, it is usually desirable to align the guide such that the index change is positive and leads to greater confinement.

With these points in mind, we use the waveguide alignment shown in Figure 2.64. The optical TE mode is launched in the <011> direction and the E-field is parallel to the $<0\bar{1}1>$ or Z' axis. The important effect for the operation of the double-Y device example is that the induced index change can be related to a phase change in the launched optical field; therefore, the field splitting to the guide with the applied dc field will undergo a phase advance relative to the field that travels through the neutral guide.

The phase change for a given length l of the guide is

$$\Delta \emptyset = \frac{2\pi}{\lambda} \Delta n l \tag{2.15}$$

The applied field in the x direction can be rewritten in terms of the applied voltage $E_x = V/t$, where V is the applied voltage and t is the guide thickness. Then

$$\Delta n' = r_{41} n_0^3 \frac{V}{2t} \tag{2.16}$$

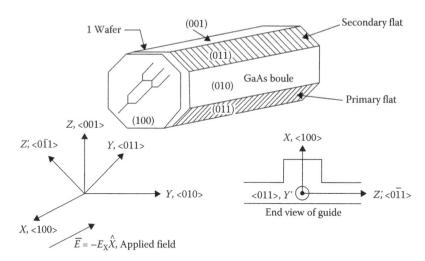

FIGURE 2.64
Crystal orientation of waveguides for the determination of electro-optic coefficients.

Substituting Equation 2.16 into Equation 2.15 yields

$$\Delta\varnothing = \frac{\pi}{\lambda} r_{41} n_0^3 V \frac{l}{t} \qquad (2.17)$$

As an example, we calculate the value of applied voltage necessary to induce a phase change of π radians for a length of waveguide l. The material parameters for a GaAs waveguide are $n_0 = 3.46$, $r_{41} = 1.2 \times 10^{-12}$ m/V. Assuming an operating wavelength of 850 nm and a guide thickness of $t = 1\,\mu m$, we find that $V = 0.018/l$ V or, given a 1 cm electrode interaction length, $V = 1.8$ V.

2.14 Nonlinear Refractive Index

The nonlinear refractive index is a third-order optical effect as opposed to the Pockels effect, which is a second-order effect. The discussion that follows is intended to explain the fundamental theoretical basis of all nonlinearities. A wide variety of third-order nonlinear optical effects is found to be related through the complete formalism of nonlinear optical susceptibilities. Among these are ellipse rotation, the "dc" Kerr effect, the optical (or "ac") Kerr effect, self-focusing, self-phase modulation, phase conjugation, and four-wave mixing.

The problem to be addressed is to solve Maxwell's equations when there are current and charge densities brought about by the interaction of light with matter (or in a vacuum where the light interacts with itself). The physical origin of these interactions is the nonlinear polarizability of either the electronic charge cloud around the nuclei or a change in the various types of nuclear motion allowed by the degrees of freedom of the material. These are referred to as either the electronic or nuclear contributions. Although the charge densities can be explained as a series of monopoles, this is found not to be applicable in the optical regime, and instead a generalized electric polarization is used. A further approximation used is the electric-dipole approximation that essentially states that the polarization is local. This in turn makes the problem independent of spatial coordinates. With this in mind, Maxwell's equations take the form

$$\nabla X E = -\frac{1}{c}\frac{\partial E}{\partial t} \tag{2.18}$$

$$\nabla X B = \frac{1}{c}\frac{\partial}{\partial t}(E + 4\pi P) + \frac{4\pi}{c}I_{dc}$$

$$\nabla \cdot (E + 4\pi P) = 0$$

$$\nabla \cdot B = 0$$

P is the local polarization and is the only time-varying source term. It is in general a function of E and fully describes the response of the medium to the field. At this point, the assumption is generally made that the electric field is sufficiently weak such that the total optical polarization density can be expanded as a power series in the electric field:

$$\vec{P} = \chi^{(1)} \cdot \vec{E} + \chi^{(2)} : \vec{E}\vec{E} + \chi^{(3)} \vdots \chi^{(3)}\vec{E}\vec{E}\vec{E} + \cdots \tag{2.18a}$$

where $\chi^{(n)}$ is the nth order complex optical susceptibility. This can be shown to be valid when the optical field is less than the atomic field, that is, when $E < E_{at} \approx 1 \times 10^9$ V/cm. A full quantum mechanical derivation dealing with the microscopic interactions in the material is necessary to exactly calculate the susceptibilities [96, 97].

The exact form of the effect observed is dependent on the frequencies of the optical fields and the polarization of their E-fields relative to the crystal axes. The frequency dependency occurs because of interactions with various resonances in the crystal. These can range from slow thermal effects (response ~1 s) to fast vibrations (response ~10^{-15} s). Drastic differences can occur as the excitation frequency moves across a resonance. This is clearly exhibited by excitonic effects in semiconductor quantum well structures.

For third-order effects, we are particularly interested in the term $\chi^{(3)}$. The real and imaginary parts of this term give rise to nonlinear refraction and absorption, respectively. The third-order contribution to the total polarization in Equation 2.18 can be written as

$$P_i^3(\omega_4) = \sum_{j,k,l} \chi_{ijkl}^3(\omega_4; \omega_1, \omega_2, \omega_3) E_j(\omega_1) E_k(\omega_2)(E_l)(\omega_3) \qquad (2.19)$$

In general, the tensor χ_{ijkl}^3 can have 81 individual elements. Fortunately, many of the elements are found to be zero or equal to each other due to various crystal symmetries. In particular, with GaAs being a $\overline{4}3$ m class crystal, the tensor reduces to 21 nonzero elements of which only 4 are independent. The frequencies of the four fields in Equation 2.19 may be of any form satisfying the conservation of energy:

$$\omega_4 = \omega_1 + \omega_2 + \omega_3 \qquad (2.20)$$

The fields of interest in the Double-Y Mach Zehnder Logic Gate, for example, are a strong pump beam of frequency ω_p, and a weaker probe beam of frequency ω_{pr}. It is found that the problem is simplified as long as the probe beam is not strong enough to affect the pump beam. Although it is not necessary for this theoretical development to make a restriction on the relationship between these frequencies, generally the beams will come from the same source material so we set $\omega_p = \omega_{pr}$. Equation 2.19 can then be written as

$$P_i^3(\omega_{pr}) = \sum_{i,j,k,l} \chi_{ijkl}^3(-\omega_{pr}; \omega_{pr}, \omega_p, -\omega_p), E_j(\omega_{pr}) E_k(\omega_p) E_i(\omega_p) \qquad (2.21)$$

The effect of interest in the Double-Y example is the alteration of the refractive index Δn by the pump beam intensity in the waveguide that in turn alters the path length for the weaker probe beam. The subsequent index can be written in the form

$$n = n_0 + \gamma \left\langle E^2 \right\rangle \qquad (2.22)$$

or

$$n = n_0 + n_2 I$$

where n_0 is the linear refractive index. It should be noted that Equation 2.22 is often written in other forms. In addition, either electrostatic units (esu) or SI units (mks) are frequently used. For this reason, we now list various conversion factors that can be used between the nonlinear coefficients [98].

$$\gamma[esu] = \left(\frac{e n_0}{40\pi}\right)n_2[m^2/W] \tag{2.23}$$

$$\gamma[cm^3/erg] = n_0(238.7)n_2[cm^2/W]$$

$$\gamma[m^2/V^2] = n_0(3.333 \times 10^{-6})n_2[cm^2/W]$$

$$\gamma[cm^3/erg] = (7.162 \times 10^7)\gamma[m^2/V^2]$$

The change in path length is usually written as a phase difference relative to a wave propagating in the other leg of the interferometer. In terms of the nonlinear index, this change can be written in the form [99]

$$\Delta\emptyset = \frac{\omega_{pr}}{c}n_2 l \tag{2.24}$$

where l is the interaction length of the pump and probe beams. It is the relationship of the nonlinear index n_2, or the index change Δn, to the third-order susceptibility that is finally developed, usually written as

$$n_2 = \frac{\Delta n}{(E^2)} \tag{2.25}$$

The last three factors to consider are the crystallographic orientation along which the pump and probe beams travel, the polarization of these beams, and the optical frequency that is used. Because it is quite lengthy to discuss the relationships for general crystallographic directions and polarizations, we give only the results for the <011> orientation and crossed beam polarizations used in the Double-Y device. As for the frequencies of the beams, two cases may be developed that are termed the resonant and nonresonant cases.

The nonresonant case is the simpler of the two and is based on the Born–Oppenheimer (BO) approximation. The assumption is made that the electrons follow the optical fields and nuclear motions adiabatically; that is, the optical frequency is much too low to interact with the electronic vibrations.

On the other hand, the frequency is still too fast to interact with the far-infrared vibrational frequencies. It is further assumed that the frequency is far from any absorption resonances. This approximation further increases the symmetry of the problem so as to reduce the number of independent elements of the susceptibility tensor to three. Generally, this is the situation when using wavelengths of lower energy than the bandgap in the Double-Y semiconductor waveguides. It is then found that the nonlinear index may be expressed [100].

$$\Delta n = \frac{12\pi}{n_0} \chi_{1122}^{(3)} (\text{esu}) \tag{2.26}$$

The second case is with the pump frequency close to the semiconductor band edge frequency of the waveguide. This is the situation when using GaAs sources with AlGaAs waveguides. In this case, a different derivation was developed by Jensen and Torabi [101] taking into account photon absorption and electron–hole recombination processes. It is found that the index variation goes from the first power dependence on the intensity, described by Equation 2.22, to a dependence of the form

$$n = n_0 + n_3 I^{1/3}$$

In this case, the nonlinear index is written as

$$n_3 I^{1/3} = \Delta n = \frac{8c_0}{3z^2 n_0 N_\gamma} \left(\alpha \frac{\tau_0}{\hbar\omega} \right)^{1/3} \tag{2.27}$$

where we use the n_3 to differentiate cases and where c_0, N_γ, α, and τ_0 are various physical constants of the material. The parameter z is a normalized frequency.

Finally, as an example, we calculate the optical power necessary for a phase change of π radians according to Equation 2.24 (assuming a 1 cm interaction length). A typical value of the nonlinear index for GaAs is $n_2 \approx 2 \times 10^{-13}$ cm²/W that indicates a required intensity of approximately 250 MW/cm² [102]. For a typical waveguide cross section of 4 μm², this indicates a required coupled input power of 10 W. Although this is an extremely high power for use in an integrated optical circuit, recent work with MQW structures suggests that the large nonlinearity observed in these materials will enhance the refractive index nonlinearity. This would conceivably allow the use of a reduced intensity while still achieving the necessary phase change. Experiment suggests intensities as low as 10^4 W/cm² may be effective, which corresponds to milliwatt power levels in the waveguides.

References

1. Midwinter, J. E. 1985. Current status of optical technology. *J. Lightwave Technol.* LT-3:927.
2. Craley, D. E., L. R. Megargel, M. A. Mentzer, and D. H. Naghski. 1987. Interconnects for VHSIC packaging. *Proc. SPIE* 85:328–336 (August).
3. Bhasin, K. B., G. Anzic, R. R. Kunath, and D. J. Connolly. 1986. Optical techniques to feed and control GaAs MMIC modules for phased array antenna applications. Paper presented at the *11th Annual Communications Conference*, March 16–20, in San Diego, CA.
4. Bhasin, K. P., G. E. Ponchak, and T. J. Kascak. 1985. Monolithic optical integrated control circuitry for GaAs MMIC-based phased arrays. *Proc. SPIE* 578 (September). (Presentation only.)
5. Mentzer, M. A. and D. E. Craley. 1987. Optical interconnects. Paper presented at the *IGK Conference on Fiber Optic Communications and Local Area Networks (FOC/LAN'87)*, October, in Anaheim, CA.
   Hutcheson, L. D. and M. A. Mentzer. 1986. Design criteria for AlGaAs integrated optoelectronic devices. *Proc. SPIE* 704 (September).
   Hutcheson, L. D. 1985. Optical interconnect technology. Paper presented at the *Annual OSA Meeting*, October 19–24, in Seattle, WA.
6. Hutcheson, L. D., P. R. Haugen, and A. Husain. 1985. Gigabit per second optical chip-to-chip interconnects. *Proceedings of the SPIE*, November, in Cannes, France.
7. Pucel, R. 1985. *Monolithic Microwave Integrated Circuits*. New York: IEEE Service Center.
8. Liao, S. Y. 1980. *Microwave Devices and Circuits*. Englewood Cliffs, NJ: Prentice-Hall, Inc.
9. Stutzman, W. L. and G. A. Thiele. 1981. *Antenna Theory and Design*. New York: John Wiley & Sons.
10. Chilton, R. H. 1987. MMIC T/R modules and applications. *Microw. J.* 30:131, 132, 134, September.
11. Johnson, R. C. and H. Jasik. 1984. *Antenna Applications Reference Guide*. New York: McGraw-Hill Book Co.
12. Brookner, E. 1987. Array radars: an update part 1. *Microw. J.* 30:117–138, February.
13. Tang, R. and R. Brown. 1987. Cost reduction techniques for phased arrays. *Microw. J.* 30:139–146, January.
14. Liao, S. Y. 1980. *Microwave devices and circuits*. New Jersey: Prentice-Hall, Inc.
15. Ibid.
16. Streetman, B. G. 1980. *Solid State Electronic Devices*. Englewood Cliffs, NJ: Prentice-Hall, Inc.
17. Wang, K. and S. Wang. 1987. State-of-the-art ion-implanted low-noise GaAs MESFET's and high-performance monolithic amplifiers. *IEEE Trans. Electron Devices* ED-34(12):2610–2615 (December).
18. Liao. *Microwave devices and circuits.*
19. Hunsperger, R. G. and M. A. Mentzer. 1988. Optical control of microwave devices: a review. *Proceedings of the SPIE Integrated Optical Circuit Engineering* VI, September, in Boston, MA.

20. Herczfeld, P. R., A. S. Daryoush, A. Rosen, P. Stabile, and V. M. Contarino. 1985. Optically controlled microwave devices and circuits. *RCA Rev.* 46:528–551 (December).

21. Seeds, A. J., J. F. Singleton, S. P. Brunt, and J. R. Forrest. 1987. The optical control of IMPATT oscillators. *IEEE J. Lightwave Technol.* LT-5 (3):403–410 (March).

    Seeds, A. J. and J. R. Forrest. 1981. Reduction of FM in IMPATT oscillators by optical illumination. *Electron. Lett.* 17(23):865–866 (November 12).

    Kiehl, R. A. 1980. Optically induced AM and FM in IMPATT diode oscillators. *IEEE Trans. Electron. Devices* ED-27(2):426–432 (February).

    Seeds, A. J. and J. R. Forrest. 1978. Initial observations of optical Injection locking of an X-band IMPATT oscillator. *Electron. Lett.* 14(25):829–830 (December 7).

    Yen, H. W. 1980. Optical injection locking of Si IMPATT oscillators. *Appl. Phys. Lett.* 38(8):630–631 (April 15).

    Forrest, J. R. and A. J. Seeds. 1978. Optical injection locking of IMPATT oscillators. *Electron. Lett.* 14(19):626–627 (September 14).

    Herczfeld, P. R., A. S. Daryoush, A. Rosen, A. K. Sharma, and V. M. Contarino. 1986. Indirect subharmonic optical injection locking of a millimeter-wave IMPATT oscillator. *IEEE Trans. Microw. Theory Tech.* MTT-34(12): 1371–1476 (December).

    Daryoush, A. S., P. R. Herczfeld, Z. Turski, and P. K. Wahl. 1986. Comparison of indirect optical injections-locking techniques of multiple X-band oscillators. *IEEE Trans. Microw. Theory Techn.* MTT-34(12):1363–1367 (December).

    Yen, H. W., M. K. Barnoski, R. G. Hunsperger, and R. T. Melville. 1977. Switching of GaAs IMPATT diode oscillator by optical illumination. *Appl. Phys. Lett.* 31(2):120–122 (July 15).

    Vyas, H. P., R. J. Gutmann, and J. M. Borrego. 1979. The effect of hole versus electron photocurrent on microwave-optical interactions in IMPATT oscillators. *IEEE Trans. Electron Devices* ED-26(3):232–234 (March).

    Schweighart, A., H. P. Vyas, J. M. Borrego, and R. J. Gutmann. 1978. Avalanche diode structures for microwave-optical interactions. *Solid State Electron.* 21:1119–1121.

22. Kiehl, R. A. 1978. Behavior and dynamics of optically controlled TRAPATT oscillators. *IEEE Trans. Electron. Devices* ED-26(6):703–710 (June).

    Gleichauf, P. H. and E. P. Eernisse. 1977. Control of TRAPATT oscillations by optically generated carriers. *IEEE Trans. Electron. Devices* 24(3):275–277 (March).

23. Salles, A. A. and J. R. Forrest. 1981. Initial observations of optical injection locking of GaAs metal semiconductor field effect transistor oscillators. *Appl. Phys. Lett.* 38:392–394 (March 1).

24. Yen, H. W. and M. K. Barnoski. 1978. Optical injection locking and switching of transistor oscillators. *Appl. Phys. Lett.* 32:182–184 (February 1).

25. Sze, S. M. 1981. *Physics of Semiconductor Devices*. New York: John Wiley & Sons, Inc.

26. Hunsperger and Mentzer. Optical control of microwave devices. A Review Proc. SPIE. 993:204–224 (1988).

27. Seeds, A. J., J. F. Singleton, S. P. Brunt, and J. R. Forrest. 1987. The optical control of IMPATT oscillators. *J. Lightwave Technol.* 5(3):403–411.

28. Seeds, A. J. and J. R. Forrest. 1981. Reduction of FM noise in IMPATT oscillators. *Electron. Lett.* 17(23):865–866.

29. Kiehl, R. A. 1980. Optically induced AM and FM. *IEEE Trans. Electron Devices* 27(2):426–432.

30. Seeds, A. J. and J. R. Forrest. 1978. Initial observations of optical injection lock-
    ing. *Electron. Lett.* 14(25):29–830.

    Yen, H. W. 1980. Optical injection locking of Si IMPATT oscillators. *Appl. Phys.
    Lett.* 36:680–683.

    Forrest, J. R. and Seeds, A. J. 1978. Optical injection locking of IMPATT oscilla-
    tors. *Electron. Lett.* 14:626–627.

    Herczfeld, P. R. et al. Indirect subharmonic optical injection locking. 1986. *IEEE
    Trans. Microw. Theory Techn.* MTT-34:1371–1376 (December).

    Daryoush, A. S. et al. 1986. Comparison of indirect optical injection-locking
    techniques. *IEEE Microw. Theory Techn.* 34(12):1363–1370.

31. Forrest, J. R. and A. J. Seeds. 1978. Optical injection locking of IMPATT oscilla-
    tors. *Electron. Lett.* 14:626–627.

32. Sze. *Physics of Semiconductor Devices.*

33. Kiehl, R. A. 1978. Behavior and dynamics of optically controlled TRAPATT
    oscillators. *IEEE. Electron Devices* ED-25:703–710.

    Gleichauf, P. H. and E. P. Eernisse. 1977. Control of TRAPATT oscillations. *IEEE
    Trans. Electron Devices* 24(2):275–277.

34. Salles, A. A. and J. R. Forrest. 1981. Initial observations of optical injection lock-
    ing. *Appl. Phys. Lett.* 38(5):392–394.

35. Yen and Barnoski. Optical injection locking and switching of transistor oscilla-
    tors. *Appl. Phys. Lett.*

36. Contarino, V. M. and A. Ortiz. A high speed phase shifter based on optical injec-
    tion. *RCA Review* 46:528–551 (December 1985).

37. Stallard, W. A., A. R. Beaumont, and R. C. Booth. 1986. Integrated optic devices for
    coherent transmission. *J. Lightwave Technol.* LT-4(7):852–857 (July).

    Alferness, R. C. 1986. Titanium diffused lithium niobate waveguide devices.
    ISAF'86 Proceedings, June 8–11 in Bethlehem, PA.

    Russ, D. 1986. The use of lithium niobate devices in optical networks. *Proceedings
    SPIE 630 Fiber Optics'86*, in London (SIRA).

    Valdmanis, J. A. 1986. High speed optical electronics: The picosecond optical oscil-
    loscope. *Solid State Technology: Test and Measurement World* S40–S44(November).

    Korotky, S. et al. 1986. Integrated optical narrow line width laser. *Appl. Phys. Lett.*
    49(1):10–12 (July 7).

    Korotky, S. 1987. Optical intensity modulation to 40 GHz using a waveguide elec-
    tro-optic switch. *Appl. Phys. Lett.* 50(23):1631–1633 (June 8).

38. Fontaine, M., A. Delage, and D. Landheer. 1986. Modeling of Ti diffusion into
    LiNbO3 using a depth dependent diffusion coefficient. *J. Appl. Phys.* 60(7):2343
    (October).

    Valdmanis, J. A. High speed optical electronics. *Solid State Technol.*

39. Russ, D. The use of lithium niobate devices. *SPIE.*

40. Ibid.

41. Young, T. P., K. K. Wong, A. C. O'Donnell, and N. J. Parsons. 1987. A compact
    LiNbO3 optical switch at 1.3 μm. *GEC J. Res.* 5(1):62–64.

    Alferness, R. C. 1986. Titanium diffused lithium niobate waveguide devices.
    ISAF'86 Proceedings, June 8–11, in Bethlehem, PA.

42. Russ. The use of lithium niobate devices. *SPIE.*

43. Alferness, R. C. Titanium diffused lithium niobate. *ISAF.*

44. Russ. The use of lithium niobate devices. *SPIE.*

45. Parsons, N. J., A. C. O'Donnell, and K. K. Wong. 1986. Design of efficient and wideband traveling-wave modulators. *Proceedings I.O.C.E.* III, April 16–18, in Innsbruck, Austria.

    Holman, R., L. Altman-Johnson, and D. Skinner. 1986. The desirability of electro-optic ferroelectric materials for guided wave devices. *Proceedings IEEE'86*, CH2358-0/86/000-0032.

    Gee, C. M. et al. 1983. 17 GHz bandwidth electro-optic modulator. *Appl. Phys. Lett.* 43(11):998–1000 (December).

    Eknoyan, O. et al. 1986. Guided-wave electro-optic modulators in Ti:-LiNbO3 at 2.6 µm. *J. Appl. Phys.* 59(8):2993–2995 (April 15).

46. Schmidt, R. V. and R. C. Alferness. 1979. Photonic switches and switch arrays on L:N60. *IEEE Trans. Circuits Syst.* CAS-26:1099.

47. Alferness, R. C. and L. L. Buhl. 1981. Waveguide electro-optic polarization transformer. *Appl. Phys. Lett.* 38:655–657.

    Donaldson, A. and K. K. Wong. 1987. Phase-matched mode convertor in LiNbO3 using near-Z-axis propagation. *Electron. Lett.* 23(25):1378–1379 (December 3).

    Alferness. R. C. Titanium diffused lithium niobate. *ISAF.*

48. Russ. The use of lithium niobate devices. *SPIE.*

49. Chung, P. S. 1986. Integrated electro-optic modulators and switches. *J. Electrical Electron. Eng.* 6(4) (December) (Australia IE Aust. and IREE Aust).

50. Wong, K. K. and R. M. DeLaRue. 1983. An improved electro-optic waveguide serrodyne frequency translator in X-cut LiNbO3 using proton-exchange. Paper presented at the *1st International Conference on Optical Fiber Sensors*, April 26–28, in London, England.

51. Stallard, W. A., A. R. Beaumont, and R. C. Booth. 1986. Integrated optic devices for coherent transmission. *J. Lightwave Technol.* LT-4(7):852–857 (July).

52. Alferness. Titanium diffused lithium niobate. *ISAF.*

53. Zang, D. and C. Tsai. 1985. Single mode waveguide microlenses and microlens arrays fabricated in LiNbO3 using titanium indiffused proton exchanged technique. *Appl. Phys. Lett.* 46:703–705.

54. Mentzer and Craley. Optical interconnects. Proceedings I6K. FOC/LAN'87.

55. Craley, D. E., L. R. Megargel, M. A. Mentzer, and D. H. Naghski. 1987. Interconnects for VHSIC packaging. *Proceedings of the SPIE 835 Integrated Optical Circuit Engineering* V, August, in San Diego, CA.

56. *LSI Logic Data Book: Bipolar and CMOS LSI/VLSI.* Dallas, TX: Texas Instruments, Inc.: 3–37.

57. *VHSIC Device Specifications Reference Book*, VHSIC Program Office (1987).

58. Ibid.

59. Grothe, H., G. Muller, and W. Harts. 1983. 560 mb/s transmission experiments using 1.3 µm InGaAsP/InP LED. *Electron. Lett.* 19(22):909–911.

60. Bowers, J. E., B. R. Hemenwaw, T. J. Bridges, and E. G. Burkhardt. 1986. Design and implementation of high-speed InGaAsP constricted mesa laser. Paper presented at the *IEE Conference on Optical Fiber Communication*, February 24, in Atlanta, GA.

61. Husain, A. 1984. Optical interconnect of digital integrated circuit and systems. *Proc. SPIE* 466:18.

62. Law, H. D., K. Hakano, and L. P. Tomasetta. 1979. State-of-the-art performance of GaAlAs/GaAs avalanche photodiodes. *Appl. Phys. Lett.* 53:180.

63. Hutcheson and Mentzer. Design criteria for AlGaAs/GaAs. Lecture Cambridge, 1983.
64. Hutcheson, Haugen, and Husain. 1987. Optical interconnects replace hardware. *J. IEEE Spectrum.* 24(3):30–35.
    Hartman, D.H., M. K. Grace, and F. V. Richard. 1986. An effective lateral fiber-optic electronic coupling and packaging technique suitable for VHSIC applications. *J. Lightwave Technol.* LT-4:73.
65. Hutcheson, L. D. 1986. GaAs/AlGaAs Monolithic optoelectronics and integrated optics. *Technical Digest of Conference on Lasers and Electro-Optics*, June 9–13, in San Francisco, CA.
66. Hutcheson and Mentzer. Design criteria for AlGaAs/GaAs. *SPIE.*
67. Ury, I., S. Margalit, M. M. Yust, and A. Yariv. 1979. Monolithic integration of an injection laser and a metal semiconductor field effect transistor. *Appl. Phys. Lett.* 34:430.
    Fukuzawa, T., N. Nakamura, M. Hiras, T. Kuroda, and J. Umeda. 1979. Monolithic integration of a GaAlAs injection laser with a Schottky gate field effect transistor. *Appl. Phys. Lett.* 36:181.
    Katz, J., N. Bar-Chaim, P. C. Chen et al. 1980. A monolithic integration of GaAs/AlGaAs bipolar transistor and heterostructure laser. *Appl. Phys. Lett.* 37:211.
    Ury, K. Lau, N. Bar-Chaim, and A. Yariv. 1982. Very high frequency GaAlAs laser field-effect transistor monolithic integrated circuit. *Appl. Phys. Lett.* 41:126.
    Kim, M., C. Hong, D. Kasemset, and R. Milano. 1983. GaAlAs/GaAs integrated optoelectronic transmitter using selective MOCVD epitaxy and planar ion implantation. *Proceedings of the IEEE GaAs IC Symposium*, October, in Monterey, CA.
68. Carney, J., M. Helix, R. Kolbas, S. Jamison, and S. Ray. 1983. Integrated optoelectronic transmitter. *Proc. SPIE* 408, April.
    Kolbas, R., J. Carney, J. Abrokwak, E. Kalweit, and M. Hitchell. 1982. Planar optical sources and detectors for monolithic integration with GaAs Metal semiconductor field-effect transistor (MESFET) electronics. *Proc. SPIE* 321, January.
    Carney, J. K., M. J. Helix, R. M. Kolbas, S. A. Jamison, and S. Ray. 1982. Monolithic optoelectronic/electronic circuits. *Proceedings of the IEEE GaAs IC Symposium*, October.
    Kim, M. E.,C. S. Hong, D. Kasemset, and R. A. Milano. 1984. GaAlAs/GaAs integrated optoelectronic transmitter. *IEEE EDL.*10.1109:306–309.
    Kilcoyne, M. K., D. Kasemset, R. Asatourian, and S. Beccue. 1986. Optical data transmission between high speed digital integrated circuit chips. *Proc. SPIE* 625:127.
69. Carney, H., R. Kolbas, S. Jamison, and S. Ray. 1983. Integrated optoelectronic transmitter. *Proc. SPIE.* 408.
70. Blauvelt, H., N. Bar-Chaim, D. Fekete, S. Margalit, and A. Yariv. 1982. AlGaAs lasers with micro-cleaved mirrors suitable for monolithic integration. *Appl. Phys. Lett.* 40:2891.
71. Carney, J., M. Helix, and R. M. Kolbas. 1983. Gigabit optoelectronic transmitter. *Proceedings of the IEEE GaAs IC Symposium*, October.
72. Walton, M. P., P. R. Haugen, and S. L. Palmquist. 1986. A 1 Gbit/s optical/electrical input monolithic GaAs transmitter IC. *Proceedings of the IEEE MTT-S International Microwave Symposium*, June, in Baltimore, MD.

73. Ray, S. and M. P. Walton. 1986. Monolithic optoelectronic receiver for Gbit operation. *Proceedings of the IEEE MTT-S International Microwave Symposium*, June, in Baltimore, MD.
74. Kilcoyne, Kasemset, Asatourian, and Beccue. Optical data transmission. *SPIE*.
75. Ibid.
76. Midwinters, J. E. and J. R. Sten. 1978. Propagation studies of graded index fiber installed on cable in operation duct route. *IEEE Trans. Commun.* COM-26 (7):1015–1020 (July).
77. Yariv, A. 1976. *Introduction to Optical Electronics,* 2nd ed. New York: Holt, Rinehart, and Winston.
78. Ross, W. E., D. Psaltis, and R. H. Anderson. 1982. Two-dimensional magneto-optic spatial light modulator for signal processing. *Proc. SPIE* 341:192.
79. Goodman, J. W., A. R. Dias, and L. M. Woody. 1978. Fully parallel, high-speed incoherent optical method for performing the discrete Fourier transform. *Opt. Lett.* 2(1):1–3.
80. Dapkus, P. D. 1982. Optical communication for IC's. Rockwell International Report, DTIC no. AD-A112239, March.
81. 1981. WDM advances enhance fiber-optic links. *EDN News*, February 20:1.
82. Schmidt, R. V. and I. P. Kaminow. 1975. Acousto-optic Bragg deflection in LiNbO3 Ti-diffused waveguides. *IEEE J. Quantum Electron.* QE-11 (1) January: 57–59.83
83. Alferness, R. C. 1982. Waveguide electro-optic modulators. *IEEE Trans. Microwave Theory Tech.* MIT-30 (8):1121–1137 (August).
84. Francis, C. L. and Instrument Development Team. 1994. Enhanced imagery through laser illumination (white paper). U.S. Army Aberdeen Test Center, Aberdeen Proving Ground, MD.
85. Nebolsine, Peter, Christopher Rollins, and Edmond Lo. 1994. High-frame rate data collection for advanced armor instrumentation enhanced imagery. Final report, subcontract no. FMD9401. Physical Sciences Inc., October.
86. Nebolsine, P., D. R. Snyder, and J. M. Grace. 2001. MHz class repetitively Q-switched, high-power ruby lasers for high-speed photographic applications. Paper presented at *AIAA Aerospace Sciences Meeting*, January, in Reno, NV.
87. Piehler, T., B. Homan, R. Ehlers, R. Lottero, and K. McNesby. 2006. High speed laser imaging, emission and temperature measurements of explosions. (ARL-RP-137). *Proceedings of the Insensitive Munitions & Energetic Materials Technology Symposium*, April, in Bristol, UK.
88. Defense Advanced Research Projects Agency Microsystems Technology Office. Super high efficiency diode sources (SHEDS) program. www.darpa.mil/mto/programs/sheds/index.html.
89. Seurin, J.-F., C. L. Ghosh, V. Khalfin, et al. 2008. High-power high-efficiency 2-D VCSEL arrays. *Proc. SPIE* 6908:690808.
90. Shaw, L. L., L. L. Steinmetz, W. C. Behrendt et al. 1984. A high-speed, eight-frame electro-optic camera with multipulsed ruby laser illuminator. Lawrence Livermore National Laboratory, UCRL-90478, September.
91. Shaw, L. L., S. A. Muelder, A. T. Rivers, J. L. Dilnaure, and R. D. Breithanpt. 1992. Electro-optic frame photography with pulsed ruby illumination. Lawrence Livermore National Laboratory, UCRL-JC-112232, November.
92. Baum, D. W., L. L. Shaw, S. C. Simonson, and K. A. Winer. 1993. Linear collapse and early jet formation in a shaped charge. *Proceedings of the14th International Symposium on Ballistics* 2, September 26–29:13–22.

93. Freitag, L., J. Kaczynkski, M. Mentzer, and D. Naghski. 2000. Packaging aspects of the LITEBUS™ parallel optoelectronic module. *Proceedings of the ECTC*, May, in Las Vegas.

94. Namba, S. 1961. Electro-optical effect of zincblende. *J. Opt. Soc. Am.* 51(1):76.
Yariv, A. 1976. *Introduction to Optical Electronics*. New York: Holt, Rinehart, and Winston.

95. Mentzer, M. A. 1982. Integrated optical logic devices. Paper presented at the U.S. *Army ARDEC Sponsored EMP/Radiation Hardening Workshop*, October, in Dover, NJ.
Mentzer, M. A., S. T. Peng, and D. H. Naghski. 1987. Optical logic gate design considerations. *Proc. SPIE* 835 *Integrated Optical Circuit Engineering* V, Mark A. Mentzer, ed., August 17–20: 362–377, in San Diego, CA.
Mentzer, M. A. 1988. Optical computing. Paper presented at the *Investigator's Meeting on Semiconductors, Optoelectronic and Magnetic Optic Materials* held at the Watertown Arsenal Materials Technology Laboratory, January, in Watertown, MA.

96. Shen, Y. R. 1984. *The Principles of Non-Linear Optics*. New York: Wiley-Interscience.

97. Hellworth, R. W. 1977. *Prog. Quant. Elect* 5(1). Pergamon Press.

98. Weber, M. J., ed. 1986. "Optical materials": Part 1. In *Handbook of Laser Science and Technology*, vol. III. Boca Raton, FL: CRC Press.

99. Hopf, F. A. and G. I. Stegeman. 1986. *Applied Classical Electrodynamics 2: Non-Linear Optics*. New York: Wiley-Interscience.

100. Hellworth. *Prog. Quant. Elect.*

101. Jensen, B. and A. Torabi. 1984. Infrared optical materials and fibers III. *Proc. SPIE* 484:159.

102. Wa, P. L. K., J. E. Sitch, N. J. Mason, J. S. Roberts, and P. N. Robson. 1985. Switching in the far-field of a nonlinear directional coupler. *Electron. Lett.* 21:26.

3

Acousto-Optics, Optical Computing, and Signal Processing

We first address the acousto-optic (AO) effect called Bragg diffraction, utilized in devices called Bragg cells. The basic operating principles of Bragg cells, the components, the applications, and the performance of some systems developed using Bragg cells will be discussed. The implementations of the AO effect are widespread. Much research continues due to the potential for great speeds and small size of optical processing devices. AO Bragg cell signal processing is an analog technology showing great promise, in an age where digital electronics is most emphasized.

3.1 Principle of Operation

Acousto-optics deals with the interaction of sound and light. In Bragg cells, the AO interaction takes place in media such as silicon, lithium niobate, tellurium dioxide, or even glasses. Figure 3.1 shows a basic Bragg cell [1]. As light enters from the left, it is interfered with by the upwardly propagating longitudinal sound waves. The sound waves are introduced by a piezoelectric transducer, which is vibrating at an RF frequency of 100 MHz to over 10 GHz. (This is the range of Bragg cell technology and not of any single Bragg cell device. Bandwidth considerations will be covered later.) The vibrations set up minute changes in the AO material's index of refraction via the photoelastic effect which creates a diffraction grating [2]. When light travels across the grating, assuming certain angular conditions, it is diffracted at an angle proportional to the RF frequency of vibration.

The angular condition is called the Bragg angle and is given by [3]

$$\sin \alpha_B = \frac{\lambda}{2\Lambda} \tag{3.1}$$

where
λ is the wavelength of the light in the acoustic medium
Λ is the acoustic wavelength

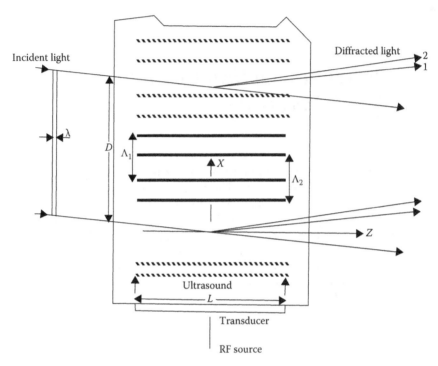

FIGURE 3.1
Bragg cell diffraction.

Note that the Bragg angle depends on the acoustic wavelength. The actual mechanism of AO diffraction is explained by a momentum diagram in Figure 3.2 [4]. The vector sum is

$$\mathbf{k_d} = \mathbf{k_i} + \mathbf{k_a} \tag{3.2}$$

where
  $\mathbf{k_a}$ is the acoustic wave vector
  $\mathbf{k_i}$ is the incident wave vector
  $\mathbf{k_d}$ is the diffracted wave vector

When the light beam enters the Bragg cell at the Bragg angle, the diffracted beam exits at the vector sum of the incident light and the acoustic wavefront. Each diffracted beam is Doppler shifted from the incident beam by the diffracting acoustic frequency. If the acoustic wavefront is composed of more than one different frequency, then the optical output will be a diffracted light beam corresponding to each input frequency. The RF frequency(s) at the transducer controls the angle at which the light beam(s) exits the Bragg cell. The properties of the Bragg cell have spurned much effort in the research

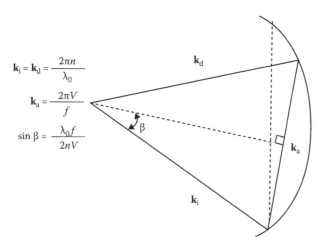

$$\mathbf{k_i} = \mathbf{k_d} = \frac{2\pi n}{\lambda_0}$$

$$\mathbf{k_a} = \frac{2\pi V}{f}$$

$$\sin \beta = \frac{\lambda_0 f}{2nV}$$

FIGURE 3.2
Wave vector diagram for AO diffraction.

and development of compact RF spectrum analyzers as well as Bragg cell correlators and convolvers for radar signal processing that are smaller and lighter than their digital VLSE and RF electronic counterparts.

3.2 Basic Bragg Cell Spectrum Analyzer

The components of a Bragg cell receiver are shown in Figure 3.3 [5]. The light source is a laser for optimum performance. The beam expander, or collimator, evenly distributes the light along the acoustic wavefront (top to bottom in Figure 3.3) to match the interaction aperture of the Bragg cell. After light is diffracted into the RF signal components a lens, called the Fourier Transform lens, focuses the light beams into a photodetector (PD) array that is mounted at the lens focal point. Each pixel of the PD array corresponds to a small frequency band, the sum of which makes up the entire bandwidth of the Bragg cell. The minimum frequency difference of two RF signals that are resolvable is approximately equal to $1/\tau$, where τ is the acoustic transit time across the interaction aperture.

The time bandwidth of the Bragg cell, $N = \tau \Delta f$, where Δf is the bandwidth of the Bragg cell, theoretically provides the total number of signals that can be simultaneously resolved by the Bragg cell receiver [6]. One virtue of an AO material is the high time bandwidth product. Figure 3.4a shows bandwidth-resolution contours for some materials [7]. The frequency bandwidth is the vertical axis, and the time delay is the horizontal axis. The time delay relates to the frequency resolution by $\Delta f = k/T$, where T is in μs, f in KHz, and k is a constant of

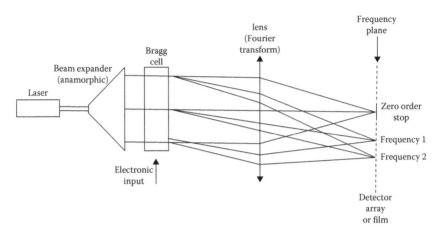

FIGURE 3.3
Basic AO spectrum analyzer.

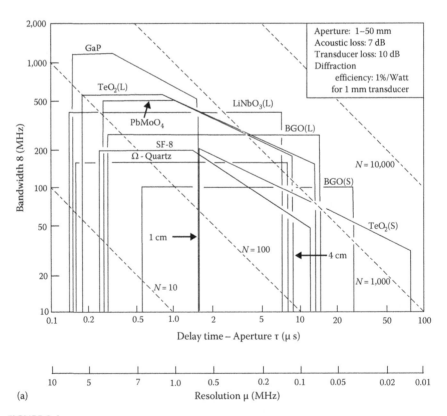

(a)

FIGURE 3.4
(a) Bandwidth-resolution contours.

Type of Cell	Characteristics
Diffraction cell	This device merely diffracts an input beam. The characteristics of the output beams are not relevant except in the cases of imaging devices. Examples include modelockers and simple modulators.
Beam modulator	This device turns an input beam "ON" and "OFF" with possible intermediate states, but it maintains beam quality under all conditions down to the "OFF" state. The primary examples would be an RF communications device.
Beam defelector	This device defelects an input beam in a controllable fashion. Beam characteristics and quality are maintained under all conditions. Examples include defelectors for laserwriting applications and cavity dumpers.
Phase modulator	This device shifts the phase of an input beam by an amount, possible variable in the "ON" state. One example is an acousto-optic delay line.
Focuser/ defocuser	This device focuses or defocuses a beam while maintaining other beam characteristics. In effect, this device uses cylindrical acoustic waves as a lens.
Frequency shifter	This device shifts the frequency of an input light beam up or down while maintaining other beam characteristics. Examples include an optical spectrum analyzer and devices for producing beat frequencies for optical information processing.
Image projector	Such a device will "break up" or transmit without distortion in a controllable fashion. As such, it can deflect or project an optical image. An example would be the projection of largescreen television images.
Information processor	Here, a variable phase grating generated by acoustic wave in a transparent medium replicates an input very-high-frequency signal. This is used to spatially modulate a light beam. Thus, an input electrical signal can be processed by spatial correlation techniques applied to a light beam transmitted through the medium.
Convert light into sound	This can happen in two related ways and is the inverse of the Debye-Sears effect. In one case, high-energy light beams create sound waves in a transparent material at their difference (beat) frequency. In other cases, pulsed high-energy light beams generate sound in a material at the pulse frequency-thermally produced sound. The sound originates from thermal expansions or contractions of the material resulting from deposition or of beam energy in the material.

FIGURE 3.4 (continued)
(b) Types of AO devices.

value of 1.2–2.0. If the bandwidth is read on the vertical axis, and the time delay on the horizontal axis, the maximum time bandwidth product of the Bragg cell can be determined. The low frequency time bandwidth product is limited by the size of the crystal that governs the acoustic wave's transit time.

The high frequency time bandwidth product is limited by the acoustic loss of the crystal. There are other parameters that govern the choice of materials, and they will be introduced later.

Some of the uses of the Bragg cell are shown in Figure 3.4b [8]. We will concentrate here on the Bragg cell receiver, especially the integrated optic Bragg cell receiver, where the beam collimator, the Bragg cell, and the Fourier Transform lens are formed on the same substrate. Figure 3.5 shows the basic IO Bragg Cell receiver [9].

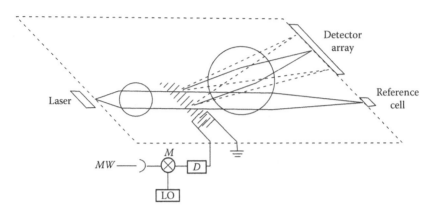

FIGURE 3.5
Integrated optic spectrum analyzer schematic.

3.2.1 Components of Bragg Cell Receivers: Light Sources

The light source is a key component in the Bragg cell receiver.

> The Bragg receiver requires a collimated, monochromatic (coherent), low-noise source for optimum performance. In the optical processor, the light source is like a local oscillator in superheterodyne receiver. Noise on the oscillator will be transferred to the output. Helium-Neon (HeNe) lasers are ideally suited for this application, possess narrow spectral bandwidths, and have demonstrated lifetimes over 10,000 hours. LED's can be used as sources in incoherent optical processors, but are not suitable for the simple Bragg cell receiver, because of their broad spectral emission. Semiconductor lasers have spectral bandwidths narrower than most LED's but not as narrow as a HeNe laser. For applications where high resolution is not required, semiconductor lasers may even improve performance because of their high output power with a much smaller size and lower power consumption. [10]

For some materials such as LiNbO$_3$, the AO diffraction efficiency using an 850 nm source can be reduced to 50% of the efficiency using a 633 nm source [11]. The higher output power of the semiconductor laser offsets this. While the choice of the source depends upon the application, for the ultimate in performance, where size and power consumption are less important, the HeNe laser is used. Where miniaturization and power consumption are most important, the semiconductor laser is the best choice. The semiconductor laser must be butt coupled to the edge of the IO Bragg cell. Materials that make a good laser, such as AlGaAs, do not possess good AO properties, and the good AO materials, such as LiNbO$_3$, cannot be made into lasers.

3.2.2 Lenses

The lenses in an IO Bragg cell are waveguide lenses. Various types of waveguide lenses have been devised, four of which are illustrated in Figure 3.6 [12]. The same effect as conventional lenses is created with step increases in thickness. A dome-shaped thickness increase is a Luneburg lens; and the inverse, a dome-shaped depression, is basically a geodesic lens. The geodesic lens is the most highly developed, because it is easier to grind the waveguide to near spherical depression than to add material to create a dome. Holographic lenses can be produced by creating grating structures with varying periodicity. The geodesic lens functions using the geometrically longer paths of rays in the center portion relative to the edges. Two lenses are used in the Bragg cell receiver—one between the light source and the cell and the other between the cell and the photodiode array.

The beam collimator determines the interaction aperture of the acoustic wavefront. The wider the aperture, the more frequency resolution is possible, to a point. The acoustic attenuation of various materials limits the aperture size.

> Several mechanisms are thought to be responsible for acoustic attenuation. In many materials, Akheiser loss caused by thermal relaxation is the dominant mechanism. For these materials, the attenuation (in dB per unit length) will increase as the square of the frequency. Other acoustic modes exhibit frequency dependencies that range from linear to square. For these materials, other mechanisms such as scattering from lattice imperfections are important in the attenuation process. The acoustic attenuation measured for certain acoustic modes tends to vary widely from sample to sample, even for pieces cut from the same crystal boule. This indicates that lattice dislocations and foreign particle impurities are dominant in the attenuation process. While measures of these crystal imperfections such as etch pit density have been correlated to optical and electrical crystal properties, these measures have not been applied in acoustic attenuation studies. [13]

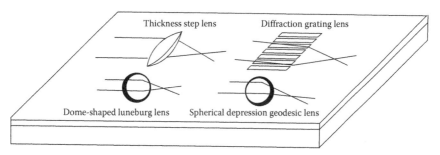

FIGURE 3.6
Four implementations of optical waveguide lenses.

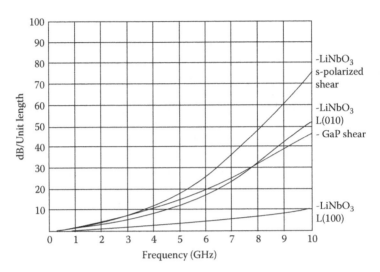

FIGURE 3.7
Acoustic attenuation in various low-attenuation AO materials.

Figure 3.7 compares several acoustic modes which exhibit low acoustic attenuation [14].

3.3 Integrated Optical Bragg Devices

Bragg cell diffraction will now be described more thoroughly. The Bragg cell diffraction efficiency (in percentage of input light deflected per watt of input RF power) is given by [15]

$$\eta = \frac{\pi^2}{2\lambda_0^3} \frac{M_3}{(f_0\tau)^{\frac{1}{2}}} \left(\frac{l}{h} \right) \tag{3.3}$$

where
    λ_0 is the optical wavelength
    f_0 is the center frequency of operation
    τ is the AO time aperture
    M_3 is the material figure of merit
    l is the normalized electrode length (and the AO interaction length)
    h is the normalized electrode height

The efficiency is stated in percentage per Watt of RF drive. Typical bulk diffraction efficiencies are 29%/W at 1.1 GHz, and 105%/W at 350 MHz, both for GaP cells.

SAW diffraction efficiencies are higher, such as 200%/W around 600 MHz, but the optical losses are higher. Equation 3.3 assumes the piezoelectric transducer is of the interdigital type that emits surface acoustic waves. The interdigital elements couple the electric excitation field directly onto a piezoelectric material. The elements are metal-deposited onto the AO material.

The light at the output of a Bragg cell is not entirely diffracted. For example, using a typical RF input power of 100 mW and the 200%/W SAW diffraction efficiency mentioned previously, the diffracted light would be 200%/W × 0.1 W = 20%, with 80% of the light undiffracted (or zeroth order). The desired light output is either of order 1, in the case of upward Doppler shifting, or −1, for downward Doppler shifting. A PD is usually put at the zeroth order focal point, after the Fourier Transform lens, to absorb energy and to act as a built-in test device for monitoring the laser output.

Referring back to Equation 3.3, it is obvious that the Bragg angle changes with the RF operating frequency. In early receiver designs, the Bragg angle was optimized for the center frequency, but the Bragg cell became very inefficient as the input frequency moved away from the center. The bandwidth of the Bragg cell is determined by the acoustic radiation pattern of the transducers. The bandwidths of modern receivers approach an octave, so there are several things that can be done to maintain good AO efficiency over a wider bandwidth. The interaction length can be decreased, but the AO efficiency also drops. More commonly, the transducers are designed as a phased array, with each element staggered forward as the light travels across the Bragg cell. Figure 3.8 illustrates this [16]. This works quite well with IO Bragg cells, especially those made with LiNbO$_3$, which is a strong piezoelectric material. The interdigital elements can easily be deposited directly onto the

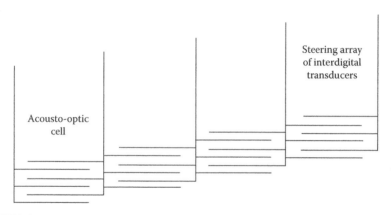

FIGURE 3.8
Staggered transducer array.

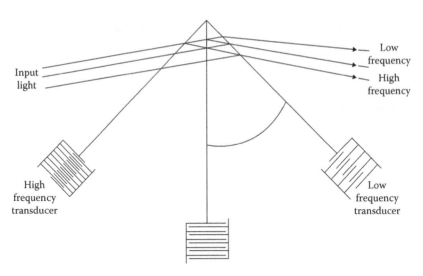

FIGURE 3.9
Angular staggered-tilted transducer array.

substrate. Another approach is the stagger-tilted array, where each element is tilted slightly to be optimized for its segment of the system bandwidth [17]. Figure 3.9 illustrates this method. The electrical impedance of a typical transducer array is about $9\,\Omega$. The impedance of the driving electronics is $50\,\Omega$, so there must be an impedance-matching transformer of some kind at the transducer interface.

3.3.1 Fourier Transform, Fourier Transform Lens

The function of the Fourier Transform lens is to rearrange the diffracted light beam output from frequency-dependent Bragg angles to frequency-dependent positions, so that the output of the PD array is a linear distribution of frequencies. The light output, or what the PDs see is three sinc functions [18]:

$$u(x_f) = k\left[\sin c\,\frac{lx_f}{\lambda F} + \frac{m}{2}\sin cl\left(\left(\frac{x_f}{\lambda F} - f\right) + \frac{m}{2}\sin cl\left(\frac{x_f}{\lambda F} + f\right)\right)\right] \qquad (3.4)$$

where
    x_f is the position of the spot
    l is the AO aperture
    m is the grating modulation depth or RF signal level
    f is the reciprocal of the diffraction grating spacing or $1/\Lambda$
    k is a constant that accounts for the laser light output and the system's optical losses
    F is the distance between the lens and the focal plane or PD array

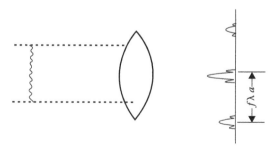

FIGURE 3.10
Fourier transform of a rectangular window with sinusoidal grating.

The first term is the zeroth order term which is not dependent on the RF input, and the last two terms are the first order terms, the "$-f$" term being the downward Doppler shift, and the "$+f$" term the upward shift. Figure 3.10 is an illustration of this.

3.3.2 Dynamic Range

The light intensity on the PD array is proportional to the square of the intensity:

$$I(x_f) = |U(x_f)|^2 \qquad (3.5)$$

This illustrates Bragg cell technology's weakness: dynamic range. For example, for a Bragg cell receiver to have 60 dB of RF dynamic range, the PD array must have 120 dB. Dynamic range will be referred to with the RF signal as the reference. One hundred and twenty decibel (60 dB RF) is beyond the performance for most PDs. The PD limits the RF dynamic range of the system to between 25 and 40 dB, depending on the type of PD used, but the Bragg cell itself has a dynamic range between 50 and 70 dB [19].

The two categories of linear array PDs useful for AO signal processing are the photodiode and the MOS depletion mode sensor [20].

> One example of a commercially available randomly addressable photodiode linear array is the Reticon CP1006 device. It consists of two interdigitated rows of 256 photodiodes with a width of 50 μm. Its dynamic range is 20 dB for noncoherent (conventional) detection processors. Another commercially available array is the Reticon CP1023 device, with CCD shift registers. It has 256 elements on an 18 μm pitch, with four independent outputs of 64 elements each, capable of a 6 MHz data rate (10.6 μs access time). Its specified dynamic range is 30 dB for noncoherent detection systems. A detector that has been built for AO Braggs cell receivers has 140 elements on a 12 μm pitch, including an element for zeroth order dumping. The measured dynamic range is 41 dB for 2 μs access time. A MOS-CCD PD has been built to be used with an IO Bragg cell receiver with four groups of 25 elements each. It is designed to be read out in four parallel channels, or two parallel channels of 50 elements.

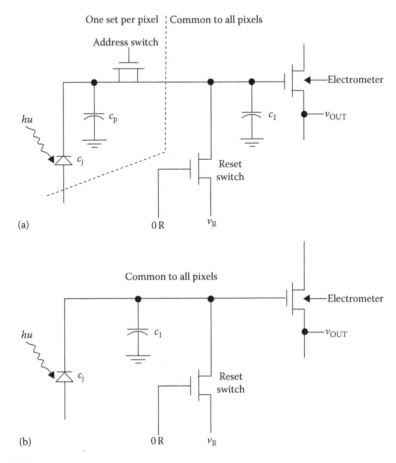

FIGURE 3.11
Schematic representation of photodiodes and CCD arrays: (a) photodiode read-out circuit; (b) CCD read-out circuit.

No dynamic range data was given. Both types of PDs are shown in Figure 3.11 [21].

3.4 Noise Characterization of Photodetectors

As detailed in the Borsuk reference [22], "the linear dynamic range of PDs for AO signal processing can be expressed as the ratio of the maximum charge capacity of the diode-capacitor combination (pel) to the rms number of noise electrons, both expressed here in units of charge:

$$D_{DET} = \frac{Q_T}{[\bar{Q}_T^2(NES)]^{\frac{1}{2}}} \tag{3.6}$$

where

Q_T is the charge capacity of a pel

NES is the noise equivalent signal

The relationship between system dynamic range and sensor dynamic range is

$$D_{SYSTEM} \leq K D_{DET} \tag{3.7}$$

where K is a function related to the system transfer function. The inequality indicates that the system dynamic range may be limited by the detector. The NES is defined as the input exposure density, E (NES), which will make the SNR equal to unity at the sensor output:

$$E(NES) = \frac{[\bar{Q}_T^2(NES)]^{\frac{1}{2}}}{R_{D_\Lambda}} \left(\frac{\mu J}{m^2}\right) \tag{3.8}$$

where

$[\bar{Q}_T^2(NES)]^{\frac{1}{2}}$ is the rms number of photoelectrons in coulombs

R_{D_Λ} is the pel responsivity given by

$$R_{D_\Lambda} = AR_\lambda = \frac{Aqn}{Hv} \tag{3.9}$$

where

H is Plank's constant

v is the optical frequency

q is the electronic charge

A is the active area of a pel

n is the total quantum efficiency

The NES is separable into two components: temporal noise and fixed pattern noise. This separation can be expressed in terms of rms noise electrons by the expression

$$[\bar{Q}_T^2(NES)]^{\frac{1}{2}} = [\bar{Q}_{temporal}^2 + \bar{Q}_{spatial}^2]^{\frac{1}{2}} \tag{3.10}$$

A summary of these noise sources is presented in the following table for MOS CCDs and photodiode arrays. The choice between selecting a photodiode

	CCD	Photodiode
Intrinsic	Bulk traps	Leakage
	Leakage	Photon shot
	Photon shot	
Circuit	Transfer inefficiency	$(kT/C)^{1/2}$ Johnson–Nyquist
	$(kT/C)^{1/2}$ Johnson–Nyquist	in diode reset operation
	in output reset operation	
		MOS electrometer
		Signal proc. Amp
		Fixed pattern
		A/D quantizing noise

and CCD for a given AO system application is dependent principally upon trade-offs between speed of operation, sensitivity, and dynamic range. In general, for low temporal noise performance ($Q_T(\text{NES})/q < 200$ e$^-$) and moderate dynamic range (10^2), CCDs are superior to photodiode arrays principally because of the low output sensing capacitance at the electrometer as opposed to the photodiode capacitance. On the other hand, for high clock rates ($\geq 5\,\text{MHz}$), sensitivity is principally limited to amplifier noise common to both CCDs and photodiode arrays to about 500–1000 e$^-$, making the higher dynamic range (10^5) obtainable with photodiodes attractive." The PD-capacitor combination responds to optical energy, so the PD response to short duration pulses, especially those that are shorter than its integration time, is a function of the pulse width. It τ is the RF ultrasonic propagation time across the Bragg cell, T_s is the integration time, and PW is the input signal pulse width, then there are three signal conditions [23]:

$$PW > T_s, \quad \tau < PW < T_s, \quad PW < T_s \qquad (3.11)$$

A graph of PW versus dynamic range is shown in Figure 3.12.

3.5 Dynamic Range Enhancement

Several optoelectronic techniques exist to extend the dynamic range of the PD. One technique utilizes two linear PD arrays and a beam splitter. Light intensity is divided unequally between the two detectors. After the photodetector which receives the majority of the light intensity saturates, the second detector array output is utilized. In this way the dynamic range of the two detector arrays can be combined to yield a total detection range which is the sum of the dynamic range of the PD arrays [24].

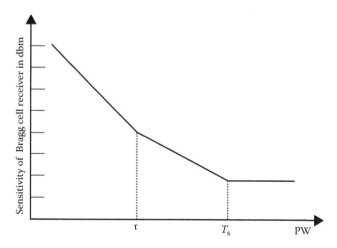

FIGURE 3.12
Sensitivity of Bragg cell receiver versus pulse width.

Another dynamic range enhancement technique uses a Mach–Zehnder interferometer (MZI) system. This approach is illustrated in Figure 3.13.

> The optical beam is split between two Bragg cells. The first of these is driven by the signals to be analyzed. The second is driven by a local oscillator reference waveform producing diffracted output reference beams which are recombined with the diffracted signal beams and then focused on the PD array. The beam recombiner is arranged to angularly shift the diffracted reference beams by a small amount so that the resultant interferometric mixing produces output signals at some intermediate frequency IF which is the same for all PD's. The heterodyned output signals are then bandpass filtered to provide immunity against DC light levels and improve discrimination against detection channels. As the PD heterodyne signal power is now proportional to the RF signal input power, the dynamic range is greatly improved [25].

This technique was accomplished using bulk techniques and has yet to be implemented using IO techniques.

3.6 Photodetector Readout Techniques

Information readout schemes embrace several different parameters. These include PD integration time, number of PDs, and pel access time. The access time and the integration time are usually matched, but a compromise must

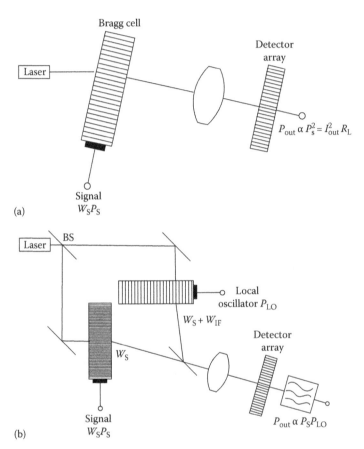

FIGURE 3.13
(a) Conventional power spectrum analyzer. (b) Interferometric spectrum analyzer.

be reached in dealing with the number of PDs and their access/integration time. It is desirable to have many PDs, but it is also desirable to access them often to detect short pulses, and to most accurately determine time-of-arrival (TOA) information in ECM applications.

For example, to achieve a 0.5 µs TOA resolution in a 1 GHz wide receiver with 1 MHz resolution (1000 PDs), a PD must be accessed every 0.5 ns. GaAs CCDs have been built that contain much of the readout circuitry on the same die (called self-scanning), but at speeds of 500 MHz, or every 2 ns [26]. Faster speed devices are in development. An alternative approach to accessing many PDs quickly is to arrange them in series–parallel readout schemes with two, four, or more PDs accessed simultaneously. More sophisticated postprocessing techniques are required, but the faster access time is worth the hardware cost if access speeds are important.

3.7 Bulk versus Integrated Optic Bragg Cells

Bulk type Bragg cell receivers now have the edge on dynamic range, bandwidth of operation, high center frequency, and frequency resolution. This is because every component of the system can be separately optimized, and the light propagates through air from component to component. IO type Bragg cells cannot have every component optimized, and compromises must be made to integrate the system. For example, the geodesic lens focuses fairly well, but a precision-ground conventional lens focuses more sharply. In IO systems, the light is coupled into a waveguide, processed, and coupled out to the PD array. The light is subject to waveguide losses of 0.5–1.0 dB cm^{-1}, coupling losses of 3 dB per coupling, and scattering caused by impurities in the waveguide. Scattering causes the PD array to have a higher background noise in frequency-adjacent pels, degrading the resolution. One advantage of IO Bragg cells is their mechanical stability and ruggedness. Everything moves together, eliminating any microphonic interference. This makes IO receivers desirable for avionic and other size-and –weight-constraint applications.

3.8 Integrated Optic Receiver Performance

Two IO Bragg cell receivers that have been developed will be discussed. The first was proposed in 1977 by Hamilton et al. [27] and in 1978 by Barnoski et al. [28]. It includes a semiconductor laser, a Y-cut LiNbO$_3$ substrate with indiffused Ti to form the waveguide, geodesic lenses, interdigital SAW transducers, and a CCD PD array coupled at the output. Design parameters were presented for a 400–800 MHz spectrum analyzer with a projected resolution of 4 MHz, and 40 dB dynamic range.

This was in response to a number of devices and system objectives. First was to identify the amplitude and frequency of emitters and to separate the different emitter categories, such as narrow pulse, wide pulse, and CW. Signals that met a specified detection criterion were to be sorted to allow reduction in data handling, while maintaining a minimum intercept uncertainty time constant. The device objectives were a monolithic integrated circuit which could interface to external systems, which operated with low noise and at high speed and low power consumption.

The first receiver built and demonstrated was by D. Mergerian et al. at Westinghouse (see Figure 3.14) [29]. A 100 µW HeNe laser was end-fire coupled into an x-cut LiNbO$_3$ crystal indiffused with Ti to 280 Å for a waveguide, and a 140 pel self-scanned photodiode array butt coupled at a 45° angle to

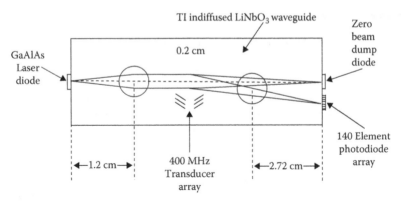

FIGURE 3.14
Westinghouse design for advanced IOSA.

the waveguide at the other end. Geodesic lenses were formed by single-point diamond turning to >0.5 μm tolerances. The lenses had an insertion loss of about 2 dB. With >5% of the input light diffracted by the SAW's, the measured dynamic range was around 40 dB. A 400 MHz bandwidth centered at 600 MHz was scanned by dividing the 140 pels into 20 groups of 7 each. A resolution of 4 MHz was achieved. The clock rate was 5 MHz, providing simultaneous pulse detection for 0.3–3.0 μs pulses separated in frequency by 20 MHz. The maximum RF power fed into the transducers was 60 mW. It is interesting to point out that the HeNe laser, emitting 100 μW, is near the power threshold of optical damage in $LiNbO_3$. The authors pointed out that performance could be greatly enhanced by using a longer wavelength semiconductor laser, due to the increasing damage threshold with increasing wavelength in $LiNbO_3$. For comparison, bulk-type receivers have been built with 1 GHz of bandwidth, 1 MHz of resolution, and 40 dB of dynamic range; however, wider bandwidth can potentially be achieved with the integrated optical spectrum analyzer using multiple arrays of tilted transducers or chirped transducers.

3.9 Nonreceiver Integrated Optic Bragg Cell Applications

Other applications of AO Bragg cells must be split into 1-D and 2-D cases, because IO implementations can only be 1-D. Only those systems realizable in IO technology will be mentioned. The space-integrating and the time-integrating correlators are shown in Figures 3.15a and b, respectively. These are 1-D devices. In the space-integrating correlator, "the received signal $g(t)$ is fed into the Bragg cell at $P1$, which is illuminated at the correct angle. Lenses L_1 and L_2 image plane P_{1a} onto plane P_{1b}. A slit filter at P_2 performs

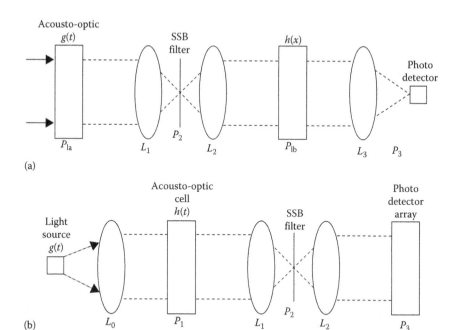

FIGURE 3.15
(a) Space-integrating AO correlators. (b) Time-integrating AO correlators.

the necessary single-sideband modulation of the data" [30] by rejecting the
zeroth- and the undesired first-order modes.

The signal incident with P_{1b} is $g(x - vt) = g(t - \tau)$. Since τ is proportional
to t, which will be the output variable, this substitution is allowable.
Thus, incident on P_{1B} is a wavefront proportional to the complex-valued
signal $g(x - \tau)$. Stored on a mask at P_{1b} is the reference transmitted signal code $h(x)$. The light distribution leaving P_{1b} is thus $h(x) = g(t - \tau)$. The
Fourier Transform of this signal is formed by L_3 at the output plane,
where we find

$$U_2(u,t) = \int g(x - \tau)h(x)e^{(-j2\tau ux)}dx \qquad (3.12)$$

When evaluated by an on-axis PD at P_3, Equation 3.12 becomes

$$U_2(t) = \int h(x)g(x - \tau)dx = h \otimes g \qquad (3.13)$$

or the correlation of g and h. The integration of Equation 3.13 is performed over distance x. The output correlation variable is time, since the
time output from the simple on-axis PD is the correlation pattern. Hence,
the name space-integrating correlator is given to this architecture [31].

Figure 3.15b shows the time-integrating system. "Here the integration is performed in time (rather than space) on the output detector. The output correlation appears as a function of distance across the output detector array. An input light source such as an LED or a laser can be modulated with the received signal $g(t)$. Lens L_1 collimates the output, and an AO cell at $P1$ is uniformly illuminated with the time-varying light distribution $g(t)$. The transmittance of the cell is now described by $h(t-\tau)$, where again $\tau = x/v$. The light distribution leaving the cell is $g(t)h(t-\tau)$. Lenses L_1 and L_2 image P_1 onto P_3 (with SSB filtering performed at P_2), where time integration on a linear PD array occurs. The P_3 light distribution (after time integration occurs on the detector) is thus

$$U_3(x) = \int h(t-\tau)g(t-\tau)dt = g \otimes h \tag{3.14}$$

or again the correlation of the received and reference signals. In this case, the integration is performed in time and the correlation is displayed in space.

The optical correlators of Figures 3.15a and b employ 1-D devices and simple imaging lenses and are relatively easy to implement. They realize the correlation operation with a moving-window transducer without the need for a matched spatial filter as in conventional correlation optical processors with fixed-format transducers. The space-integrating system can accommodate large range-delay searches between the received and reference signals; however, the signal integration time and time-bandwidth product that this system can handle are small, limited by the aperture (40 μs dwell time is typical) and TBWP (1000 is typical) of the AO cell.

In the time-integrating system, the signals must be time aligned, and only a much smaller range-delay search window (equal to 40 μs typical aperture time of the cell) is possible; however, the time-integrating processor allows longer integration time (limited by the integration time and noise level of the detector) and the associated correlation of longer TBWP signals" [32].

3.10 Optical Logic Gates

The AO signal processing systems discussed previously may be considered representative of the traditional analog optical computing technology. Technological niches for analog optical computing include image enhancement and noise reduction, spectrum analysis of RF signals, pattern recognition, and signal correlation in radar, sonar, and guidance systems. Another class of devices is directed toward the development of digital optical computer systems. Combinations of only 10–20 of these gates could potentially improve a hybrid opto-electronic processor speed by a factor of 100–1000. The following examples are representative of this emerging device technology.

3.10.1 Introduction

All optical signal processing systems for both optical communications and optical computing have tremendous potential for dramatically improving the information handling capacity and data rate in various signal processing systems by keeping the information in optical form during most of the processing path [33]. Major improvements in real-time signal analysis can be obtained through the implementation of all-optical circuits due to the advantage of low power consumption, high speed, and noise immunity. Additionally, all optical systems lend themselves directly to implementation in highly parallel, high throughput architectures.

Dynamic nonlinear optical processes lie at the heart of these optical logic devices. The index of refraction of a semiconductor can be significantly varied through the creation of electrons, holes, and excitons or by exposure to high intensities of light. Changes in these properties can induce absorptive or dispersive bistability in the semiconductor system [34]. A system is said to be optically bistable if it has two output states for the same value of the input over some range of input values. Relaxation properties of the excited semiconductor state dictate the characteristics of these optically induced optical changes.

Gibbs et al. [35] in 1979 were among the first to apply excitonic effects to optical bistability in GaAs. Bistability was demonstrated just below the exciton resonance in a cryogenically cooled, $4\,\mu m$ thick, GaAs sample utilizing approximately $200\,mW$ of optical power. The device exhibited a switch-off time of approximately $40\,ns$. This bistability was due to the change in refractive index resulting from nonlinear absorption in the vicinity of the exciton resonance. One limitation to further progress in this area was that excitonic resonances are generally not seen in room-temperature direct-gap semiconductors due to temperature broadening effects.

It has been found, however, that multiple quantum well (MQW) structures show strong exciton resonances at room temperature due to the QW confinement and consequential increase in binding energy of the excitons [36]. These resonances show saturation behavior similar to that observed at low temperature in conventional semiconductors. As such, the MQW structure represents a preeminent class of device to be researched for implementation in optical logic function units.

Bulk and integrated bistable devices have demonstrated the ability to perform a number of logic functions [37]; however, various device parameters such as the physical size, power requirements, and speed and temperature of operation have generally proved to be impractical. Two devices that show promise as optical processing components are the Double-Y

Mach–Zehnder Interferometric Optical Logic Gate (Double-Y) and the MQW Oscillator.

3.10.2 Interferometer and Quantum Well Devices

Individual MZIs have been demonstrated in various materials and have been used as modulators and switches [38]. The MQW Oscillator makes use of the nonlinear optical effects unique to the MQW structure. Both the Double-Y, which is an extension of the MZI concept, and the MQWO devices lend themselves directly to fabrication in GaAs material. Both devices can overcome the limitations of speed and operational temperature mentioned previously. In its present configuration, the Double-Y is about 2 cm long and requires rather large optical powers for switching. On the other hand, the MQW is already a small, low-power device.

The Double-Y and MQW oscillator devices utilize a total of three optical effects to perform their logic and switching functions. These are the linear electro-optic, or Pockels effect, the nonlinear index of refraction, and the Quantum Confined Stark Effect (QCSE). Derivations of the first two effects are included at the end of Chapter 2. The Double-Y device has been studied and fabricated in LiNbO$_3$ [39] and in AlGaAs [40]. The MQW Oscillator can be implemented in several forms, one of which has been studied by Wood et al. [41].

A schematic diagram of the Double-Y device is shown in Figure 3.16. The signal of interest is input at the center or control leg. The signal splits at the Y and then recombines. The output depends on how this signal recombines. If the path lengths down the two legs are identical (or differ by an integral number of wavelengths) then the control signal recombines in phase and is referred to as a "one." If the path lengths differ by an integral number of half wavelengths, the signal recombines destructively, resulting in a "zero" output.

The Double-Y device utilizes two nonlinear optical effects to change the path length of either or both legs and thus change the output state between "one" and "zero." The Pockels effect provides a phase shift proportional to an electric field applied across the waveguide. This field is supplied by applying a voltage to the electrodes over the waveguide. The circuit is completed with a ground plane below the guide. These contacts are indicated in Figure 3.16. In the Double-Y device, this effect is used to tune the recombination of the signal. Tuning is needed to correct for any differences in the physical lengths of the two legs and to set the initial output state of the device to either "one" or "zero." Calculations indicate that for the geometry shown in Figure 3.16, a phase change of 180° can be obtained by applying a potential of approximately 1.8 V.

The other effect used to change the path length is the nonlinear or intensity dependent index of refraction. By coupling a high intensity of light into either of the signal legs, "A" or "B," the path length of the control pulse will be changed. It is seen, therefore, that after setting the initial state by using

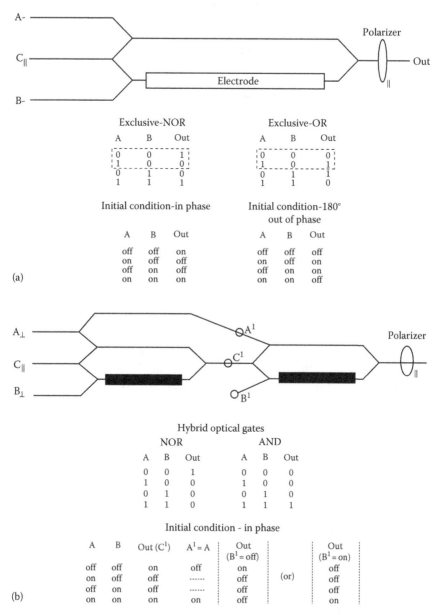

FIGURE 3.16
(a) "Double-Y" Mach–Zehnder interferometer logic gate. (b) Cascade MZI gate.

(*continued*)

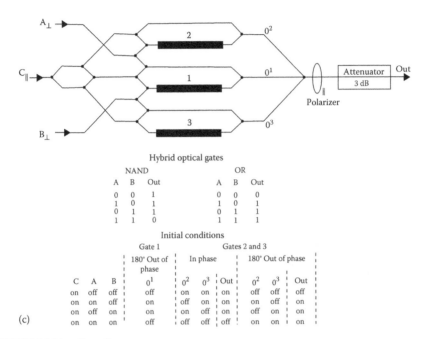

FIGURE 3.16 (continued)

(c) Hybrid MZI gate.

the Pockels effect, the relationship between the inputs "A" and "B" and the output will be exclusive—or (XOR) or inverse XOR logic functions, as described by the truth tables in Figure 3.16. One should note that the output polarizer selects only the parallel polarized control signal for the output. This has implications for the concept of cascading the basic Double-Y gate to implement other logic functions.

Several combinations of cascaded Double-Y type gates can be considered in order to implement all basic logic functions. Review of Figure 3.16a shows that three logic functions are accounted for by the basic Double-Y gate. Two of these are the $\overline{\text{XOR}}$ and XOR functions. In addition, it can be seen that by using the gate in either of the above initial phase modes (i.e., in-phase or 180° out-of-phase) an inversion of the "A" input can be obtained by leaving the "B" input "off" or "on," respectively. This is indicated by the dashed boxes in the truth tables.

Figure 3.16b shows a cascaded gate that produces either the NOR or AND function, depending on the setting of the $B_\perp$ input. A few of the possible problems that could be encountered in the operation of this gate are the splitting of the A input power and the length of the $A_\perp$ guide. Both of these factors can make the proper phase changes and synchronization difficult to obtain. These problems can be overcome with proper design control. Another problem is the extended length of the device; however, this would be reduced in accordance with reductions in the length of the single Double-Y gate.

Figure 3.16c shows the implementation of the OR and NAND functions. Although structurally more complex than the previous gates, it is a very symmetric design; and therefore, synchronization should be easier to obtain. On the other hand, the effects of the various splitters, combiners, and cross-overs will need to be optimized.

A major problem with cascading the Double-Y logic gate is that an "on" output from a following gate cannot be generated by inversion of an "off" output from a previous gate. This is because the individual gates inherently use "on" control pulses. As these are progressively canceled, subsequent gates have fewer operations that can be performed. This is markedly different from electrical logic gates where a source to invert an "off" value is always available.

The MQW Oscillator of interest for optical computing is a hybrid device, which relies on an electrical signal to change the absorptive properties of the material. This change in absorption, caused by the QCSE [42] is a very important effect in MQW structures. These structures are crystals grown by alternating very thin layers of two materials which have similar lattice constants but different band gap energies. A schematic diagram of the material structure and the band diagram of a typical MQW structure is shown in Figure 3.17. The interesting quantum effect arises from the confinement of excitons in the narrow QWs of the material. Confining these excitons, which normally have a radius of about 300 Å, produces two effects. The first is that the exciton absorption line, which normally broadens into the band gap absorption with increasing temperature, remains resolvable even at room temperature. This effect is illustrated in Figure 3.18. The second effect is an induced anisotropy in the effect of an applied electrical field. The exciton absorption line is shifted by an electric field applied perpendicular to the structure layers. This is due to the QW's enhancement of the exciton binding

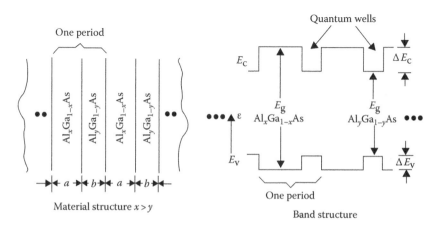

FIGURE 3.17
Multiple quantum well structure.

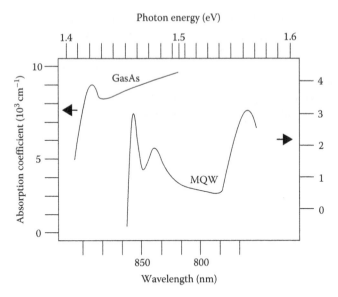

FIGURE 3.18
Linear absorption spectra of room-temperature GaAs and MQW samples. (After Miller, D.A.B. et al., *Appl. Phys. Lett.*, 41(8), 679, 1982.)

and consequent avoidance of the exciton dissociation until the higher than normal field strengths. These effects are illustrated in the Figure 3.19.

3.11 Quantum Well Oscillators

Considerable potential exists for applications of the exciton resonance in MQWS. MQWS are constructed from alternating layers of GaAs and $Al_xGa_{1-x}As$. When an optical wave is sent through an MQWS, an optical absorption peak which initially creates excitons can be observed near the optical band edge. Excitons are particles described by wave packets made up of electron conduction band states and hole valence band states.

There are two different types of excitons that exist in MQWS. One is an electron-light hole exciton and the other is an electron-heavy hole exciton. Whenever we refer to a "hole" it can be either a heavy hole or a light hole. The excitons that are trapped in the QWs have discrete energy levels that can be described quantum mechanically as the energy levels of a "particle in a box."

An induced electric field perpendicular to the MQWS layers tends to reduce these energy levels, shifting the absorption peaks to lower frequencies. This shifting effect is nonlinear and smaller for smaller fields. This behavior is

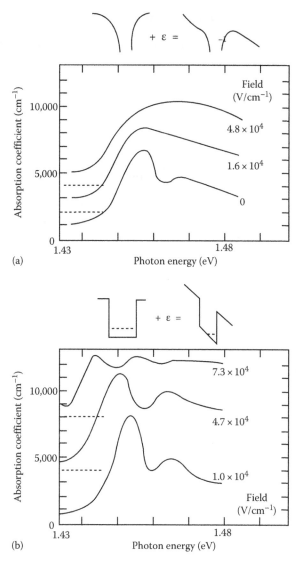

FIGURE 3.19
Absorption spectra at various electric fields for applied (a) parallel, (b) perpendicular, to the
QW layer. The distortion is shown schematically. The zeros, indicated by dashed lines, are
displaced for clarity. (After Chemla, D.S. et al., *J. Opt. Soc. Am.*, 2(7), 1162, 1985.)

known as the QCSC and is analogous to the Franz–Keldysh effect or band-
edge shift observed in bulk semiconductors. The shift in exciton energy levels is described as a function of perpendicular electric field intensity. The
method used to describe the energy level of excitons in MQWS is to calculate
the energy of one exciton in a single QW [41].

3.11.1 Description of the Quantum Well

The QW is formed by the discontinuities in the energy gap between the GaAs and the $Al_xGa_{1-x}As$ layers. The energy gap difference is described by

$$\Delta E_g(eV) = 1.425x - 0.9x^2 + 1.1x^3$$

where x is the mole fraction of aluminum. This energy gap difference is divided between the hole QWs and the electron QWs. Two theories exist regarding the magnitude of this QW split. The first is an 85:15 split, which means that 85% of the energy gap difference is at the conduction band discontinuity and 15% is at the valence band discontinuity. Another theory suggests a 57:43 split. It is interesting to note that the use of either one of these theories results in little or no difference in the magnitude of the energy shift. Figure 3.17 shows a schematic diagram of a QW structure indicating the band discontinuities.

To describe the exciton's energy levels in a QW, we must first know the Hamiltonian. The Hamiltonian for this problem can be written as the sum of terms due to the contribution from the individual electron and hole, and the interaction between the two particles.

$$H = H_e + H_h + H_{ex} \tag{3.15}$$

where

$$H_e = H_{KEz_e} + V_e(z_e) - eF_{\perp}z_e$$

$$H_h = H_{KEz_h} + V_h(z_h) - eF_{\perp}z_h$$

$$H_{ex} = H_{KEr_{eh}} + V_{e-h}(r, z_e, z_h)$$

The z direction is defined perpendicular to the MQWS layers and z_e and z_h are the positions of the electron and hole, while r is the relative position of the electron and hole. The kinetic energy operators in the z direction are defined by

$$H_{KEz_e} = \frac{-\hbar^2}{2m_{e\perp x}^4} \frac{d^2}{dz_e^2} \tag{3.16}$$

$$H_{KEz_h} = \frac{-\hbar^2}{2m_{h\perp z}^k} \frac{d^2}{dz_h^2}$$

where

$m_{e\perp x}^k$ and $m_{h\perp z}^k$ are the effective masses of the electron and hole in the z direction

$V_e(z_e)$ and $V_h(z_h)$ are the depths of the electron and hole QW potentials

$F_\perp$ is the electric field perpendicular to the MQWS layers

The effective masses for both of the discontinuity split ratios can be calculated as shown in Figure 3.20.

The kinetic energy operator for the interacting particles or exciton and their reduced effective mass are

$$H_{\text{KEr}_{eh}} = \frac{-\hbar^2}{2\mu} \frac{d^2}{dr^2} \qquad (3.17)$$

$$\mu = \frac{m_{e\parallel}^* \; m_{h\parallel}^*}{m_{e\parallel}^* + m_{h\parallel}^*}$$

with $m_{e\parallel}^*$ and $m_{h\parallel}^*$ being the effective masses of the electron and hole in the plane parallel to the layers. There are the normal bulk effective masses. The Coulomb potential energy due to the interaction of the electron and hole is

$$V_{e-h}(r, z_e, z_h) = \frac{-e^2}{\varepsilon(|z_e - z_h|^2 + r^2)^{1/2}} \qquad (3.18)$$

3.11.2 Solution of the Exciton Energy

The next step to finding the energy levels is to compute the expectation value of H. This gives us the energy of the exciton

$$E_{ex} = \langle \Psi | H | \Psi \rangle \qquad (3.19)$$

or

$$E_{ex} = E_e + E_h + E_b$$

	85:15	57:43
Me^*	$0.0665 + 0.0835x$	$0.0665 + 0.0835x$
M_{h^*-h}	$0.45 + 0.31x$	$0.34 + 0.42x$
M_{l^*-h}	$0.088 + 0.049x$	$0.094 + 0.043x$
V_e (meV)	340	228
V_h (meV)	60	172

FIGURE 3.20
Effective mass calculation table.

where

$$E_e = \langle \Psi | H_e | \Psi \rangle$$

$$E_h = \langle \Psi | H_h | \Psi \rangle$$

$$E_E = \langle \Psi | H_{ex} | \Psi \rangle$$

E_e and E_h are the energies of the electron and hole respectively, which can be found using Equation 3.15. E_B is the exciton-binding energy which is found using Equation 3.15.

The wave function Ψ is written in a separable form as

$$\Psi = \psi_e(z_e)\psi_h(z_h)\varnothing_{e-h}(r) \tag{3.20}$$

where

$$\varnothing_{e-h}(r) = \left(\frac{2}{\pi}\right)^{1/2}\frac{1}{\lambda}e^{[-r/\lambda]}$$

The parameter λ is used as a variational parameter describing the amplitude radius of the exciton.

3.11.3 Determination of E_e and E_h

The problem of determining the energies of the electron and the hole in the QW is subject to a great deal of simplification. Initially it may be noted that, due to the separable nature of the wave function Ψ, Equations 3.19 reduce to the following:

$$E_e = \langle \psi_e | H_e | \psi_e \rangle \tag{3.21}$$

$$E_h = \langle \psi_h | H_h | \psi_h \rangle$$

The next simplification is that a wave function describing an electron and hole in an infinite QW can be substituted for the wave function ψ_e or ψ_h if an effective well width L_x is used in place of the actual QW width. The longer well width effectively accounts for the fact that there is significant penetration of the wave function into the QW barriers.

The Schrödinger equation for the infinite well barrier in a uniform perpendicular electric field is

$$\frac{-\hbar^2}{2m_\perp^*}\frac{d^2}{dz^2}\varsigma(z) - (W + eF_\perp z)\varsigma(z) = 0 \tag{3.22}$$

where $\varsigma(z)$ is either the electron or hole wave function. Note that the parameters $m_{\perp}^{*}$ and e and the dimensions z and L_{x} all depend on the particle being investigated.

By making the substitution

$$Z = \left[\frac{2m_{\perp}^{*}}{(e\hbar F_{\perp})^{2}} \right]^{1/3} (W + eF_{\perp}z) \tag{3.23}$$

we can transform Equation 3.22 into an Airy differential equation:

$$\frac{d}{dZ^{2}} \varsigma(z) - Z\varsigma(z) = 0 \tag{3.24}$$

The solutions have the form

$$\varsigma(z) = bAi(Z) + cBi(Z) \tag{3.25}$$

where b and c are constants and $Ai(Z)$ and $Bi(Z)$ are Airy functions defined by

$$Ai(Z) = c_{1}f(Z) - c_{2}g(Z)$$

$$Bi(Z) = \sqrt{3}[c_{1}f(Z) - c_{2}g(Z)]$$

Further definitions are

$$f(Z) = 1 + \sum_{n=1}^{\infty} \frac{Z^{3n}}{(3n)(3n-1)(3n-3)(3n-4)\cdots 3\cdot 2}$$

$$g(Z) = Z + \sum_{n=1}^{\infty} \frac{Z^{3n+1}}{(3n+1)(3n)(3n-2)(3n-3)\cdots 4\cdot 3}$$

while $c_{1} = 0.35503$ and $c_{2} = 0.25882$.

It is convenient to work with dimensionless parameters for the energy and the field. These are defined as

$$w = \frac{W}{W_{1}}; \quad f = \frac{F}{F_{1}}$$

The energy W_1 is taken to be the lowest energy level of the particle with zero field. This is simply

$$W_1 = \frac{\hbar^2}{2m_\perp^*} \left[\frac{\pi}{L_x} \right]^2 \qquad (3.26)$$

The value W_1 is also used to define the unit for the electric field

$$F_1 = \frac{W_2}{eL_x}$$

The values of Z at the QW walls are

$$Z_\pm = Z(\pm \tfrac{1}{2} L_x)$$

and therefore, using Equations 3.23 and 3.26 $Z_\pm$ becomes

$$Z_\pm = \left[\frac{-\pi}{f} \right]^{1/3} (W + \tfrac{1}{2} f) \qquad (3.27)$$

If we use the boundary conditions that $\varsigma(Z)_\pm = 0$ at the QW walls we find from Equation 3.25 a resulting equation which completely determines the eigen energy:

$$0 = Ai(Z_+)Bi(Z_-) - Ai(Z_-)Bi(Z_+)$$

This equation is completely independent of well width and effective mass.

3.11.4 Determination of E_B

The exciton-binding energy E_B can be rewritten in the following form

$$E_B = E_{KE_r} + E_{pE_r} \qquad (3.28)$$

E_{KE_r} is the kinetic energy of the relative electron–hole motion in the layer plane. It is described by the following equation:

$$E_{KE_r} = \langle \varnothing_{e-h}(r) | H_{KE_{reh}} | \varnothing_{e-h}(r) \rangle = \frac{\hbar^2}{2\mu\lambda^2} \qquad (3.29)$$

E_{pE_r} is the Coulomb potential energy of the electron–hole relative motion.

$$E_{pE_r} = \langle \Psi | V_{e-h} | \Psi \rangle$$

In this case, we must use variational wave functions for ψ_e and ψ_h in Equation 3.20 [43]. These are written as

$$\psi_{e,h} = N(\beta)\cos\left[\frac{\pi z}{L_x}\right]\exp\left[-\beta\left[\frac{z}{L_x}+\frac{1}{2}\right]\right] \tag{3.30}$$

Note that β (β_e or β_h), L_x (L_e or L_h) and z (z_e or z_h) all depend on whether we are using ψ_e or ψ_h. $N(\beta)$ is a normalization function and is defined such that

$$N^2(\beta) = \frac{4\beta}{L_x[1-\exp(-2\beta)]}\frac{\beta^2+n^2}{\beta^2}$$

β is a variational parameter and is calculated by minimizing the function $E(\beta)$ with respect to β^2

$$E(\beta) = E_1^{(0)}\left[1+\frac{\beta^2}{4\pi^2}+\chi\left[\frac{1}{\beta}+\frac{2\beta}{4\pi^2+\beta^2}-\frac{1}{2}\coth\left(\frac{\beta}{2}\right)\right]\right] \tag{3.31}$$

where the ground state energy at zero field is

$$E_1^{(0)} = \frac{\hbar^2\pi^2}{2m_\perp^* L_x^2}$$

and the dimensionless electro-static energy is

$$\chi = \frac{|e|F_\perp L_x}{E_1^{(0)}}$$

This is calculated separately for both the electron and hole.
Using Equations 3.18, 3.20, and 3.30 Equation 3.20 becomes

$$E_{pE_r} = \frac{-2e^2}{\pi\varepsilon\lambda^2}N^2(\beta_e)N^2(\beta_h)\int_{\theta=0}^{2\pi}\int_{r=0}^{\infty}\int_{z_e=-L_e/2}^{L_e/2}\int_{z_h=-L_h/2}^{+L_h/2}\cos^2\frac{\pi z_e}{L_e}\exp\left[-2\beta_e\left[\frac{z_e}{L_e}+\frac{1}{2}\right]\right]$$

$$X \cos^2 \frac{\pi z_h}{L_h} \exp\left[-2\beta_h\left[\frac{z_h}{L_h}+\frac{1}{2}\right]\right] X \frac{r\exp(-2r/\lambda)}{[(z_e-z_h)^2+r^2]^{1/2}} \, d\theta \, dr \, dz_e \, dz_h \quad (3.32)$$

The integral over θ is trivial and the integrals over z_e and z_h must be done using numerical methods. We can calculate the integral over r using the following equation:

$$G(t) = \frac{2}{\lambda} \int\limits_{r=0}^{\infty} \frac{r\exp\left(\dfrac{2r}{\lambda}\right) dr}{\sqrt{t^2+r^2}} = \frac{2|t|}{\lambda}\left[\frac{\pi}{2}\left[H_1\left(\frac{2|t|}{\lambda}\right)-N_1\left(\frac{2|t|}{\lambda}\right)-1\right]\right] \quad (3.33)$$

where
    $H_1(u)$ is the first-order Struve function
    $N_1(s)$ is the first-order Neumann function or Bessel function of the second
    kind

The Struve function is defined by [44]

$$H_1(u) = \frac{2}{\pi}\left[\frac{u^2}{1^2\cdot 3}-\frac{u^4}{1^2\cdot 3^2\cdot 5}+\frac{u^6}{1^2\cdot 3^2\cdot 5^2\cdot 7}-\cdots\right]$$

while the Neumann function is defined by [45]

$$N_1(s) = \frac{\pi}{2}\pi Y_1(s)+(\ln 2-\varphi)J_1(s)$$

where φ = Euler's constant,

$$Y_1(s) = \frac{-2}{s\pi}+\frac{2}{\pi}\ln\left(\frac{s}{2}\right)J_1(s)-\frac{s}{2\pi}\left[\sum_{k=0}^{\infty}\{\xi(k+1)+\xi(k+2)\}\frac{(-s^2/4)^k}{k!(1+k)}\right] \quad (3.34)$$

$$J_1(s) = (s/2)\sum_{k=0}^{\infty}\frac{(-s^2/4)^k}{k!(1+k)}$$

$$\xi(1) = -\varphi$$

$$\xi(n) = -\varphi+\sum_{k=1}^{n-1} k^{-1} \quad n\geq 2$$

In summary, to calculate E_b for a particular filed we must first deduce β_e and β_h for the field, using a variational calculation, and then calculate E_{KE_r} and E_{PE_r}, adjusting λ variationally to minimize E_B.

3.11.5 Effective Well Width Calculations and Computer Simulation

In order to perform all the previous calculations, we must substitute an effective QW width for the actual well width. Note that for an infinite well the wave function has a value of zero outside the well walls. Because the wells in the MQWS are finite, the effective well widths for the infinite well model must be greater than the actual well widths to account for the significant penetration of the wave function into the finite QW barriers. There are two basic ways to find the effective well widths. One way is to try to match the wave functions of the finite well as closely as possible with the infinite well. Another way is to equate the energy of the first energy level in the finite well with the infinite well.

3.11.5.1 Matching of Wave Functions

In order to understand how we match the wave functions, Figure 3.21 shows a finite well wave function where

$$\Psi_{finite} = \Psi_I + \Psi_{II} + \Psi_{III}$$

$$\Psi_I = 2B\cos(\kappa a)e^{k(z+\alpha)} \tag{3.35}$$

$$\Psi_{II} = 2B\cos(\kappa z)$$

$$\Psi_{III} = 2B\cos(\kappa a)e^{k(z-a)}$$

$$\kappa a\tan(\kappa a) = ka$$

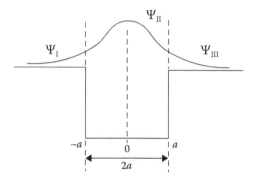

FIGURE 3.21
Finite well wave function.

$$\kappa a^2 + (ka)^2 = \frac{2m_\perp^* a^2 |V_z|}{\hbar^2}$$

B is a normalization constant and a is the well half width. Note that the values of $m_\perp^h$ and V_z depend on whether the calculation is for an electron or hole.

Figure 3.22 shows the infinite well wave function where

$$\Psi_{infinite} = \left(\frac{2}{L_x}\right)^{1/2} \cos\left(\frac{\pi z}{L_x}\right)$$

L_x is the effective well width and depends on the particle type used in equation.

Figure 3.23 shows both wave functions superimposed. In this figure, the shaded region is the difference in area between the finite and infinite wave functions. The area difference is expressed by the integral

$$\text{Area difference} = \int_{-\infty}^{\infty} (\Psi_{finite} - \Psi_{infinite})^2 dx \qquad (3.36)$$

$$= \int_{-\infty}^{\infty} (\Psi_{finite}^2 - 2\Psi_{finite}\Psi_{finite} + \Psi_{infinite}^2) dx$$

$$= 2 - 2\int_{-L_z/2}^{+L_z/2} \Psi_{finite}\Psi_{infinite}\, dx$$

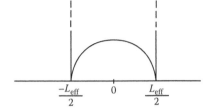

FIGURE 3.22
Infinite well wave function.

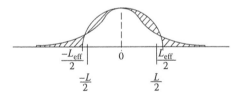

FIGURE 3.23
Infinite and finite well wave functions.

To match these functions as closely as possible, the area between the functions must be minimized.

3.11.5.2 Equating Energy

An alternative method of calculating L_x is to equate the energies of the first energy levels and solve for the length of the infinite QW. The following simple formula is the result:

$$L_x = \frac{\pi}{\kappa}$$

where κ is determined from the finite wave function by Equation 3.35.

3.11.5.3 Computer Calculations

In order to complete the calculations described above, computer calculations were used. The first step in calculating the effective width is to find the effective masses using the equations in Figure 3.20. Additionally, the height of the potential well barrier must be calculated using the energy gap difference. The split ratio must be taken into account when finding these two values. By using equations and, the value of κ and k can be found. From these two values, B can be calculated using the fact that

$$\int_{-\infty}^{\infty} |\Psi_{finite}|^2 \, dx = 1$$

At this point we can calculate the effective width using the equivalent energy method (Section 3.11.5.2) because we know the value of κ. Assuming we have completed the integration in Equation 3.36, we have sufficient information to find the effective width that minimizes the area difference between the two wave functions. The computer can find this effective width iteratively, using Equation 3.36 with different effective well widths until the minimum is found. The values we obtained for several well widths and aluminum concentrations are listed in Figure 3.24.

3.11.6 Summary of Calculation Procedure

With the full theory developed in the previous sections, a brief summary is helpful to explain the steps needed to obtain the exciton energy. Equation 3.19 shows that the exciton energy is broken down into three energy values. The quantum well is fully specified by what may be considered as three "fundamental" values. The aluminum concentrations of the barrier and the well material determine the band gap difference. The other parameter is the actual width of the well.

57:43 Split

Well Width (Å)	Aluminum Concentration	L_{eff} electron (Å)		L_{eff} heavy hole (Å)		L_{eff} light hole (Å)	
		$\psi_{inf} = E_{fin}$	$E_{inf} = E_{fin}$	$\psi_{inf} = \psi_{fin}$	$E_{inf} = E_{fin}$	$\psi_{inf} = \psi_{fin}$	$E_{inf} = E_{fin}$
90	0.1	169.5	144.7	122.5	116.8	170	144.9
	0.3	125	118.5	106.5	104.3	128	120.5
95	0.1	172.5	149.5	127.5	121.8	173	149.7
	0.3	129.5	123.5	111	109.3	132.5	125.4
100	0.1	176	154.2	132	127	176.5	154.5
	0.3	134.5	128.4	116	114.3	137	130.4

85:15 Split

Well Width (Å)	Aluminum Concentration	L_{eff} electron (Å)		L_{eff} heavy hole (Å)		L_{eff} light hole (Å)	
		$\psi_{inf} = \psi_{fin}$	$E_{inf} = E_{fin}$	$\psi_{inf} = \psi_{fin}$	$E_{inf} = E_{fin}$	$\psi_{inf} = \psi_{fin}$	$E_{inf} = E_{fin}$
90	0.1	150	134	145	121.3	283	191
	0.3	117.5	113.2	117	112.7	169	144.4
95	0.1	154	139	149	136	281	195
	0.3	122.5	118.2	121.5	117.6	172	149.2
100	0.1	157.5	143.8	153	141	279.5	200
	0.3	127	123.2	126.5	122.6	175.5	154

FIGURE 3.24
Effective widths table.

Step A—The effective masses of the particles, the conduction–valence band split, and the effective well widths are found as follows:

A1: The energy gap difference is used to determine the depths of the conduction and the valence band wells. There are the values V_e and V_h. A relatively arbitrary choice is made concerning the ratio of the conductance-valence band split.

A2: Figure 3.20 is used to determine the effective masses of the electrons, and light and heavy holes.

A3: Using the values determined in steps A1 and A2, along with the actual width of the wells, the effective well widths are determined by the methods previously described.

Step B—Values of E_e and E_h are found using the following steps.

B1: A "universal" curve is calculated using Equation 3.27. It is convenient to keep these values as a look-up table since the solution requires an iterative technique. This curve expresses the normalized energy w as a function of the normalized filed f through Equation 3.27.

B2: With a knowledge of $w(f)$ the actual energy of the particle W (which is actually E_e or E_h) is found by use of the Equation 3.26 to produce W1, F1, f, w, and W.

Step C—Finally, the binding energy E_B is determined. For each value of electric field $F_\perp$:

C1: Determine β_e and β_h for Equation 3.30. This is done by minimizing Equation 3.31 with respect to β.

C2: Using a value of λ and the above values of $N^2(\beta)$ calculate the value of E_{PE_r} by performing the double integration over z_e and z_h in Equation 3.32.

C3: Calculate E_{EE_r} using Equation 3.29.

C4: Repeat steps C2 and C3, adjusting λ to minimize E_B.

C5: Start at step C1 for the next field value.

Again, it is useful to use look-up tables for parts of Step C—in particular for the values of Equation 3.33.

3.11.7 Example: Fabrication of MQW Oscillator

An MQW could be fabricated from GaAs/AlGaAs heterostructure such as that shown in Figure 3.25. This structure would require about five mask and fabrication steps. The thin optical window, with no field applied perpendicular to the QW junction layers, will be transparent to a wavelength—denoted λ_{low}—of slightly lower energy than the absorption peak. Application of a filed will shift the absorption peak to lower energies and the window will become opaque to the incident wavelength λ_{low}. This operation, depicted in Figure 3.26 [41] shows the transmission of the device as a function of the applied bias.

The proposed vertical material structure is shown in Figure 3.27 [46]. Figure 3.28 shows a photoluminescence scan of the material taken at room temperature (~300 K). Various important features and their wavelengths and

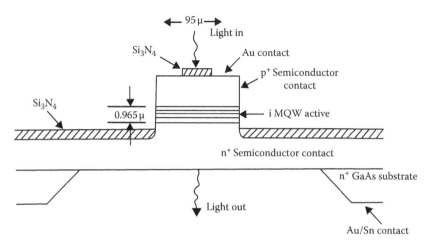

FIGURE 3.25
Optical QW oscillator modulator.

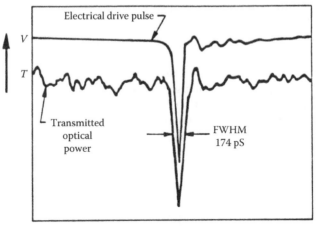

MQWO modulator response

FIGURE 3.26
MQWO modulator response.

Layer		t	
8	p-AIGaAs 1×10^{18}	9800 Å	
7	20 Periods Ⓐ p—1×10^{18}	1940 Å	
6	30 Periods Ⓐ undoped	2910 Å	
5	50 Periods Ⓑ undoped	9500 Å	
4	30 Periods Ⓐ undoped	2910 Å	
3	20 Periods Ⓐ n—5×10^{17}	1940 Å	
2	n-AIGaAs 5×10^{17}	9800 Å	
1	n-GaAs 5×10^{17}	2000 Å	
Substrate	n$^+$ GaAs		

Ⓐ

GaAs	28.5 Å
AlGaAs	68.5 Å

Ⓑ

GaAs	95 Å
AlGaAs	95 Å

Note:
AlGaAs = Al$_{.3}$Ga$_{.7}$As

FIGURE 3.27
MQWO device structure.

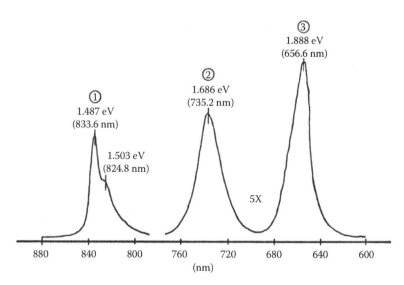

FIGURE 3.28
Photoluminescence scan of MQWO device at room temperature.

corresponding photon energies are indicated in the figure. Most notable are the obviously well-defined exciton emissions, which would not be observable at room temperature without the QW structures. Peak 1 indicates the exciton from the equal width QW structure, denoted period B in figure. Peak 2 shows the emission from the unequal, period A wells. Peak 3 indicates emission from the wide band gap AlGaAs p-type cap layer (layer 8 in Figure 3.27). The energy of emission corresponds to an aluminum concentration of approximately 32%. Figure 3.29 shows another photoluminescence

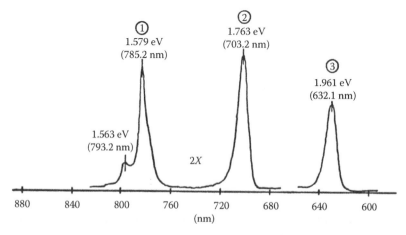

FIGURE 3.29
Photoluminescence scan of MQWO device at 12 K.

scan performed on the sample at a temperature of 12 K. Two effects are noted. First is the shift of the peaks. The second is the sharpening of the peaks due to the decrease of the phonon energy broadening.

3.12 Design Example: Optically Addressed High-Speed, Nonvolatile, Radiation-Hardened Digital Magnetic Memory

Military and space applications require a number of unique memory characteristics. These include [47] nonvolatility, maintainability, security, low power and weight, small physical size, radiation hardness, and the ability to operate in severe environments (temperature, humidity, shock and vibration, and electromagnetic fields). The need for radiation hardness is of special interest where the use of various directed energy weapons (DEW) against satellite assets is anticipated. The goal of DEW is to destroy at least part of the threat nuclear warheads in the exoatmosphere. This scenario will expose the on-board electronics and associated computer memory to many types of high-level radiations.

Considering these requirements, along with the goals of further improved life cycle cost, programmability, speed, volume, and power consumption of present devices—the crosstie memory system discussed herein represents an excellent choice for nonvolatile memory in a radiation environment. The design discussion increases the read speed of this memory by more than an order of magnitude, providing the possibility for multiple applications.

3.12.1 History of the Magnetic Crosstie Memory

The crosstie magnetic effect in thin film magnetic material was discovered in 1958 by E.E. Huber. The magnetic thin film exists in two possible magnetic states: the Neel wall state which represents the fundamental magnetic domain wall; and the crosstie/Bloch line pair on the Neel wall, which represents the second magnetic domain state. The crosstie wall constitutes a transition form between Bloch walls of very thick films and the Neel walls of very thin films. As the film thickness is decreased, the Neel wall forms segments of opposite polarity, which reduces the magnetostatic energy in the wall. These segments are separated by perpendicular magnetization circles called Bloch lines, which minimize energy by forming spiked walls or crossties on alternate Bloch line locations. The crosstie Bloch line pairs represent digital information stored in the computer memory.

Work began in the early 1970s to develop a computer memory utilizing serrated strips of thin film 80–20 Ni–Fe permalloy data tracks that stored information in a series of crosstie Bloch line patterns on a Neel wall. The concept was similar to the serial propagation of bubbles in a magnetic bubble memory, but promised faster performance. The serial crosstie memory presented numerous development problems, however, illustrating the need for an alternative approach [48]. The random access approach was successfully demonstrated in early 1982 by Mentzer, Schwee et al. [49]. Its unique characteristics as a faster, nonvolatile, radiation- and temperature-insensitive, random access device make the crosstie memory a viable alternative for many applications. More recently [50], this technology is being developed for alternative flash memory, fast-start computer systems, biomedical storage devices, and dynamic RAM replacement. Significant efforts are underway as of this writing at IBM, Infineon, Naval Research Laboratory, and elsewhere.

3.12.2 Fabrication and Operation of the Crosstie Memory

Thin film permalloy (81–19 Ni–Fe) patterns on a silicon substrate support stable magnetization domain states which can be switched rapidly and detected nondestructively. The geometry of the basic memory cell is shown in Figure 3.30 [51]. The contours represent the direction of magnetization in the memory cell. The cell on the left represents a digital zero (Neel wall); and the cell on the right represents a digital one state (crosstie Bloch pair on Neel wall). The crosstie wall constitutes a transition state between Bloch walls of thick films and Neel walls of very thin films. The fundamental cell geometry supports two states—the Neel wall and the Bloch line-crosstie pair on the Neel wall.

Vacuum deposition of the permalloy alloy is performed in a magnetic field resulting in approximately 400 Å isotropic films with $H_k < 10$ Oe.

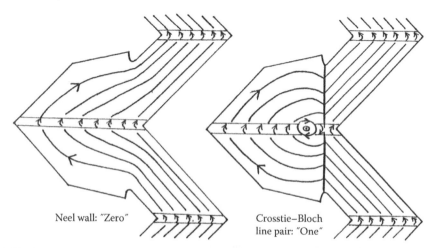

FIGURE 3.30
Magnetic domain pattern in CRAM storage element.

Silicon substrate temperature and alloy stoichiometry are such that the material displays zero magnetostriction and maximum magnetoresistance $\Delta R/R > 2\%$. This ratio indicates the differential resistance between fields applied parallel and perpendicular to permalloy current flow, divided by the permalloy resistance. The current method of detecting the memory state utilizes this effect.

Switching between digital information states is accomplished through application of coincident in-plane fields (~10 Oe) produced by 5–7 mA currents flowing through overlying row and column conductors. The magnetic film changes states when a sufficient current is placed in the local area. Therefore, the column conductors zig-zag or meander in order to align the local electromagnetic fields of the column conductor with the row conductor in the local area of the memory cell. The test chip described by Mentzer et al. [52] (Figure 3.31) consisted of four 64-bit columns, a fifth reference column, column meander columns, column meander conductors, and four row conductors, providing access to 16 test bits. Five vacuum depositions and five photolithographic steps are required to fabricate the device. The permalloy memory cells are followed by a sputtered silicon nitride dielectric layer with vias for contacts to the permalloy, metallization for row conductors and contacts to the permalloy, another dielectric layer, and metallization for column conductors.

The electronic method for nondestructive read-out (NDRO) is performed by applying a positive field less than the annihilation level using the appropriate row conductor. The permalloy thin film has current applied. Therefore, when the current in the row conductor passes through a memory

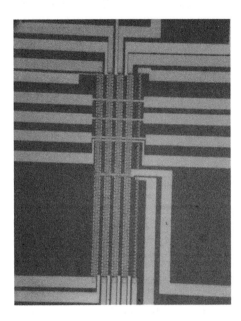

FIGURE 3.31
Photomicrograph of 4 × 64 crosstie memory array.

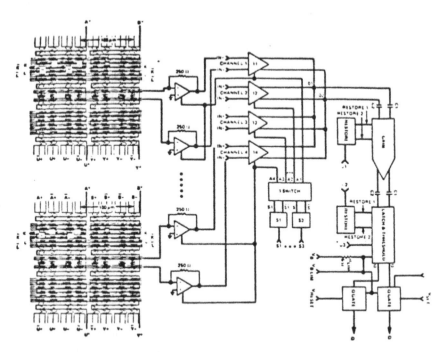

FIGURE 3.32
Signal extraction and processing architecture for CRAM.

cell location in a crosstie state, a differential voltage is created. The differential detection voltage for a single bit obtained when a field is applied in the annihilation direction but less than the annihilation threshold is less than 300 μv. This low-detection signal level represents a significant challenge to the widespread implementation of the crosstie memory. As a result, Mentzer et al. [53] devised a signal extraction and processing architecture (Figure 3.32), which improved the signal level but increased the complexity of the device significantly.

The Mentzer/Lampe technique utilizes double-correlated sampling to remove offset nonuniformity fixed patterns by means of serial subtraction. Postdetection integration is implemented where signals accumulate constructively and directly while temporal noise is uncorrelated and adds only in terms of power. Integration of four such samples effectively doubles the signal to noise ratio. This approach also severely reduces the achievable memory density and increases power consumption, since the dominant factor dictating both density and power consumption is the size of the current drivers required to supply current through the permalloy columns. Clearly a technique is needed whereby the detection scheme is improved, in favor of a more direct readout—but still utilizing all of the useful features of the memory device (electronic write, radiation hardness, etc.).

3.12.3 Potential Optical Detection Scheme

Optical detection could supplant the magnetoresistance detection approach, which limits access times to approximately 500 ns. This is too slow for many primary memory applications. The optical technique will afford read-write cycle times that are competitive with the fastest available technologies on the market. In fact, the crosstie memory access times may prove to be the fastest achievable, with the optical addressing scheme described herein.

The detection circuit for the optical detection (read) scheme requires simpler circuitry; so there will be not only performance gain but also significant improvement in cost and producibility. Additional limitations of the magnetoresistance effect include the requirement to heat the substrate to increase crystallite size and to provide the necessary magnetoresistance ratios, along with temperature limitations. These limitations are eliminated with optical access and detection using the Faraday magneto-optic effect.

Magneto-optic Faraday rotation occurs when linearly polarized light is transmitted through a material parallel to the direction of a magnetic field in the material. The amount of rotation φ is proportional to the magnetic field M and the path length t, such that $\varphi = VHt \cos \theta$, where V is the Verdet constant (rotation per unit length per unit magnetic field), and θ is the angle between the magnetic field and direction of light propagation. In the proposed crosstie memory detection configuration, the presence or absence of a crosstie wall is used to produce a change in the transmitted optical power density through crossed polarizers that are rotated from extinction by an angle equal to the amount of Faraday rotation produced. Differences in transmitted light intensity through different memory locations thus provide detection of stored information. The Neel wall state does not have a magnetic field perpendicular to the permalloy surface while the crosstie state does.

Figure 3.33 shows the permalloy pattern to be used to write information on the memory device. Creation of crossties is performed by the application of coincident currents flowing through conductor strips insulated by nitride layers. Windows should be added to the nitride and conductor layers to define the spot size of the transmitted light for optical detection of the crossties. Polarizers can be either grown or bonded to the device as wafer processing permits. Laser radiation will be applied through a particular memory block or array accessed via electro-optic switching.

Detection of transmitted light intensity is accomplished with a silicon PD array. Random access or parallel addressed read may also be accomplished through the detector array, rather than electro-optic switching. Current levels produced by changing light intensities resulting from partial extinction of Faraday rotated light are amplified and coupled to a comparator with sensitivity controls allowing identification of logic levels. The device configuration is illustrated at Figure 3.34.

Assuming a crosstie magnetic field equal to the saturation magnetization, we have a specific Faraday rotation in permalloy equal to $1-2 \times 10^5$ deg cm^{-1}.

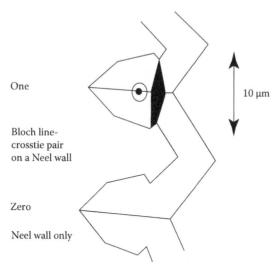

One

Bloch line-
crosstie pair
on a Neel wall

10 µm

Zero

Neel wall only

FIGURE 3.33
Permalloy array pattern.

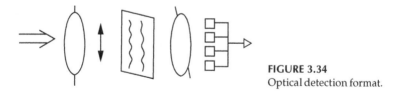

FIGURE 3.34
Optical detection format.

This provides a rotation of approximately 2° with a permalloy thickness of 600 Å, at $\lambda_0 = 830$ nm. Thin film and multilayer dielectric polarizers supply a 3.5% change in transmitted intensity per degree of analyzer rotation with the polarizer at 45° from extinction with respect to the input polarization. A fixed insertion loss of 15% for the polarizers, and a loss due to absorption in the permalloy of 2.7% for 600 Å thickness ($\alpha = 6 \times 10^5$ cm^{-1}) along with a 50% loss due to initial polarizer rotation, provide a decrease in optical power density form polarizer to detector of 65%. This level is decreased by 3.5% polarizer transmittance per degree of rotation X2 = 7% due to the presence of a crosstie. If we couple 3 mW of power into the system, the PDs will see changes of 74 µW on a background of 1050 µW.

PDs can produce 25 mV µW^{-1} µs^{-1} integration time at 12 µm center-to-center spacings. A power level of 1050 µW then provides a 650 mV signal using 25 ns integration time, with a 46 mV change in voltage for crosstie detection. Excellent signal to noise ratios are thus possible for low false alarm probabilities. The signal may be compared to a reference zero location electro-optically switched simultaneously, and differentially amplified for further signal processing. Such a system provides excellent high-speed detection for the crosstie memory, and eliminates the problems associated with electrically addressed readout.

3.12.4 Radiation Hardening Considerations

Radiation environments in which such a memory may be required to survive are a strong function of the particular system deployment. Tactical man-operated systems are required to survive only modest ionizing dose rates, low accumulated doses and low neutron levels, with man being the limiting factor. Strategic ground-based systems are required to withstand sharply higher levels. Space-based systems with no nuclear weapon environment must survive only the natural Van Allen belt energetic electron and proton environments, which may range up to several tens of kilorads (SiO) depending on the orbit and the duration of the mission. Military systems in space must survive the nuclear weapon environments of high ionizing dose rates, high accumulated doses and neutron fluences. The system memory requirements in terms of organization, power, speed, etc., will all be different in each of these applications; but in general will tend toward higher speed, low power, high density, nonvolatility, and low soft error rates.

The optical read crosstie memory discussed affords all these features in addition to nuclear hardness. Since the optically read memory is inherently free from soft errors, there will be no need to increase the word length to allow for error detection and correction. Normally, 5 bits of a 16 bit word are utilized for double error, single error correction; so an immediate savings of 30% of the memory size and power supply requirement is realized. The radiation environments in which memories may be required to survive vary with application, but may include ionizing dose rates in excess of 10^{12} rad (Si)/s, total accumulated doses of up to 10^7 rad (Si), and neutron fluences of up to 10^{14} n cm^{-2}.

References

1. Oakley, W. S. October 1979. Acousto-optic processing opens new vistas in surveillance warning receivers. *Defense Electron.* 11:91–101.
2. For the relations between index change and acoustic power, see: Pinnow, D. 1970. Guidelines for the selection of acoustooptic materials. *IEEE J. Quant. Electron.* 6:223–238.
3. Korpel, A. January 1981. Acousto-optics—A review of fundamentals. *IEEE Proc.*, Vol. 69(1):48–53.
4. Chang, I. C. March 1986. Acousto-optic channelized receiver. *Microwave J.* 29:142, 142–144.
5. Oakley, W. S. Acousto-optic processing opens new vistas. *Defense Electron.*

6. Chang, I. C. 1986. Acousto-optic channelized receiver. *Microwave J.* 29:141–144.
7. Tsui, J. B. 1983. *Microwave Receivers and Related Components*, Chap. 7, Avionics Laboratory, Air Force Wright Aeronautical Laboratories. NTIS.
8. Main, R. P. Fundamentals of acousto-optics. *Laser Appl.* June 1984:111.
9. Lawrence, M. November 1984. Acousto-optics may replace electronic processing. *Microwaves RF.*
10. Oakley, W. S. Acoustooptic processing opens new vistas. *Defense Electron.*
11. Mellis, J., G. R. Adams, and K. D. Ward. February 1986. High dynamic range interferometric Bragg cell spectrum analyser. *IEEE Proc. J.* 133(1):26–30.
12. Hamilton, M. C. January–February 1981. Wideband acousto-optic receiver techniques. *J. Electron. Defense.*
13. Kalman, R. and I. C. Chang. 1986. Acousto-optic Bragg cells at microwave frequencies. *Proceedings of SPIE 639.*
14. Ibid.
15. Ibid.
16. Lawrence, M.J., B. Willke, M.E. Husman, E.K. Gustafson, and R.C. Byer. 1999. Dynamic response of a Fabry-Perot interferometer. *J. Opt. Soc. Am.* B. 16(4):523–532.
17. Tsui, J. B. 1983. *Microwave Receivers and Related Components.* Chapter 1. Avionics Laboratory Air Force Wright Aeronautical Laboratories. NTIS.
18. Ibid.
19. Oakley, W. S. Acousto-optic processing opens new vistas. *Defense Electron.*
20. Borsuk, G. M. January 1981. Photodetectors for acousto-optic signal processors. *IEEE Proc.* 69(1):100–118.
21. Ibid.
22. Ibid.
23. Tsui, J. B. *Microwave Receivers and Related Components.* Avionics Laboratory.
24. Borsuk, G. M. 1981. Photodetectors for acousto-optic signal processors. *IEEE.* 69(100):100–118.
25. Mellis, J., G. R. Adams, and K. D.Ward. 1986. High dynamic range interferometric Bragg cell. *IEEE Proc. J.* (1):26–30.
26. Borsuk, G. M. Photodetectors for acousto-optic signal processors. *IEEE.*
27. Hamilton, M. C., D. A. Wille, and W. J. Miceli. 1977. *Opt. Eng.* 16:467.
28. Barnoski, M. K., B. Chen, H. M Gerard, E. Marom, O.G. Ramer, W.R. Smith Jr., G.L. Tangonan, R.D. Weiglein. 1978. Design, fabrication and integration of components for an integrated optic spectrum analyzer. *Paper Presented at the IEEE Ultrasonics Symposium. IEEE Trans. on Sonics and Ultrasonics.* 26(2):146.
29. Mergerian, D., E. C. Malarkey, R. P. Patienus et al. September 1980. Operational integrated optical RF spectrum analyzer. *Appl. Optics* 19(18):15.
    Mergerian, D., E. C. Malarkey, M. A. Mentzer. 1982. Advanced integrated optic RF spectrum analyzer. *Proceedings of SPIE. Conference on Integrated Optics II,* January 28–29, Los Angeles, CA. 321:149.
30. Casasent, D. June Short course at Carnegie Mellon, 1981. Optical signal processing. *Electro Optical Systems Design.*
31. Ibid.
32. Ibid.
33. Mentzer, M. A., S. T. Peng, and D. H. Naghski. 1987. Optical logic gate design considerations. *Proceedings of SPIE 835 Integrated Optical Circuit Engineering V,* Mark A. Mentzer, ed., August 17–20, pp. 362–377, San Diego, CA.

34. Gibbs, H. M., S. L. McCall, and T. N. C. Venkatesan. 1980. Optical bistability. *Opt. Eng.* 19:463.
35. Gibbs, H. M., S. L. McCall, T. N. C. Venkatesan, A. C. Gossard, A. Passner, and W. Wiegmann. 1979. Nonlinear optical materials. *Appl. Phys. Lett.* 35:451.
36. Esaki, L. and R. Tsu, 1970. Superlattice and negative differential conductivity in semiconductors. *IBM J. Res. Dev.* 14:61–65.
37. Gibbs, H. M., S. L. McCall, T. N. C. Venkatesan. 1976. Differential gain and bistability using a sodium-filled Fabry-Perot interferometer. *Phys. Rev. Lett.* 36:1135.
    Smith, P. W., E. H. Turner, and P. J. Maloney. 1978. Electro-optic nonlinear Fabry-Perot devices. *IEEE J. Quant. Electron.* QE-14:207.
    Jewell, J. L., Y. H. Lee, M. Warren. 1985. GaAs étalon optical logic devices. *Appl. Phys. Lett.* 46:918.
38. Martin, W. E. 1975. Refractive index profile optimization. *Appl. Phys. Lett.* 26:562.
    Ramaswami, V., M. D. Divino, and R. D. Stanley. 1978. *Appl. Phys. Lett.* 32:644.
39. Lattes, A., H. A. Haus, F. J. Leonberger, and E. P. Ippen. 1983. An ultrafast all-optical gate. *IEEE J. Quant. Electron.* QE-19 (11):1718.
    Haus, H. A., N. A. Whitaker, Jr., and M. C. Gabriel. 1985. All-optical logic devices using group III–IV semiconductor waveguides. *Proceedings of SPIE 578 Integrated Optical Circuit Engineering III*, p. 122.
40. Mentzer, M. A., S. T. Peng, and D. H. Naghski. Optical logic gate design considerations. *SPIE.*
41. Wood, T. H., C. A. Burr, D. A. B. Miller, D. S. Chemla, and T. C. Damen. 1985. *IEEE J. Quant. Electron.* QE-21 (2):117.
42. Mentzer, M. A., S. T. Peng, and D. H. Naghski. Optical logic gate design considerations. *SPIE*, pp. 362–377.
43. Bastard, G., E. Mendez, L. H. Chang, and L. Esaki. 1983. Variational calculations on a quantum well in an electric field. *Phys. Rev. B* 28 (6):3241.
44. Abromowitz, M. and I. Stegan, eds. 1972. *Handbook of Mathematical Functions with Formulas and Graphs.* New York: John Wiley and Sons, p. 496.
45. Ibid., p. 360.
46. Miller, D. A. B., D. S. Chemla, D. J. Eilenberger, D. J. Smith, P. W. Gossard, and W. T. Tsang. 1982. Optical bistability due to increasing absorption. *Appl. Phys. Lett.* 41(8):679.
47. Lo. D. S. Nov. 1984. Selecting memory devices for military applications. *Defense Electron.*
48. Schwee, L. J., P. E. Hunter, F. A. Restorff, and M. T. Shepard. 1982. CRAM Concepts and initial studies. *J. Appl. Phys.* 53(3):2762.
49. Mentzer, M. A., C. W. Baugh, E. A. Hubbard. 1982. Fabrication and characterization of a crosstie random access memory. *IEEE Trans. Magnetics*, 3IM3 Conference, July, Montreal, Quebec.
50. Slaughter, J. M. August 2009. Materials for magnetoresistive random access memory. *Annu. Rev. Mater. Res.* 39:277–296.
51. Mentzer, M. A., C. W. Baugh. 1982. Magnetic crosstie random access memory. Best Paper Award. *Government Microcircuit Applications Conference*, Orlando, FL.
52. Ibid.
53. Mentzer, M. A., D. Lampe. 1984. Signal processing and architecture for the crosstie random access memory. *Presented at Government Microcircuits Applications Conference*, Las Vegas, NV.

4

Fiber-Optic Sensors

4.1 Introduction

Fiber-optic sensing is a relatively new area of technology in which a number of systems are currently in development. The sensing mechanisms in a fiber sensor can include one or all of the following modulation techniques: intensity, phase, polarization, or frequency. The high sensitivity of fiber sensors makes them attractive in many applications where electronic sensors are inadequate due to limitations in sensitivity, range, and resolution. The interferometric class of fiber sensors offers the highest sensitivity and dynamic range. The parameters to be sensed cause a change in the phase of the interferometer, eliminating the need for a separate modulator [1].

The general advantages of fiber sensors include the following: fiber communication system compatibility, large bandwidth/high data rate, high sensitivity and dynamic range, EMI immunity, and geometric versatility [2]. A wide range of physical parameters can be detected with the fiber sensor, as evidenced by the following representative fiber sensor types [3]: hydrophone (pressure), magnetometer (magnetic field), gyroscope (rotation rate), position sensor (displacement), pressure sensor, vibration sensor (acceleration), flow sensor, liquid-level sensor, pollution monitor, temperature sensor, and pH sensor.

4.2 Amplitude Modulation Sensors

Fiber sensors are often separated into two broad categories: amplitude modulation (or intensity sensor) and phase modulation sensor. The intensity sensor simply monitors the interruption of light propagation through a (normally multimode) optical fiber due to the perturbing field to be measured. A transduction device or material transfers the influence of the measured field into a disruption or interruption of the optical fiber path, modulating the intensity of light transmitted to a detector. Signal processing at the detector may be used to measure frequency information related to the perturbing field or to separate out the effect of interest. This type of sensor is illustrated in block diagram format in Figure 4.1.

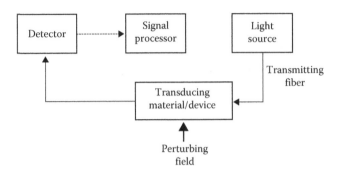

FIGURE 4.1
Intensity sensor.

Fiber waveguide can also be used simply to transmit information from any transducer, including electromechanical or electronic, to a detection or processing point. These sensors are the more fundamental of the fiber sensors, and use multimode commercially available components in their construction. Perhaps of more significant interest in terms of performance that is superior to electronic sensors are the interferometric or phase modulation sensors.

4.3 Phase Modulation Sensors

In the phase sensor, the perturbing field to be detected modulates, via a transducer, the phase of coherent light in the sensing arm of an interferometer. Figure 4.2 illustrates the fiber interferometric, or phase sensor. A frequency-stable laser light source is coupled to a single-mode optical fiber and then split into two legs of the sensor system: a sensing leg and a reference

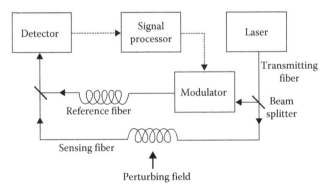

FIGURE 4.2
Phase sensor.

leg. The detector output is used to modulate the reference leg to maintain a constant-phase relationship to the sensing leg. In this approach, the level of feedback is a measure of the sensing leg perturbation.

Such a constant-phase modulation and detection scheme is commonly employed because, in many cases, the output sensitivity is more sensitive when the system is biased at quadrature. This sensitivity factor is illustrated in Figure 4.3. With an additional 90° phase difference between the two sensor legs, the phase variation to be detected produces a larger detector output variation than without bias. Movements as small as 10^{-13}m can be detected with the bulk and fiber configurations of these sensors [4].

Perturbing fields detected in this manner can include the following: acoustic, electric, magnetic, acceleration, trace, vapor, pressure, temperature, rotation, and electric current. Common applications for such sensors are hydrophones, magnetometers, gyroscopes, and accelerometers. In the discussion to follow, the fiber magnetometer is discussed as representative of the class of interferometric sensors.

Before discussing more details of the sensor design, it is instructive to look at the major classes of interferometers in their bulk configurations and then to visualize these sensors in a fiber configuration [5]. Figure 4.4 illustrates the four most common bulk interferometer configurations: the Michelson, Mach–Zehnder, SAGNAC, and Fabry–Perot.

In the Michelson configuration, the reference path is from the laser, off the beam splitter, to the fixed mirror, and through the beam splitter to the

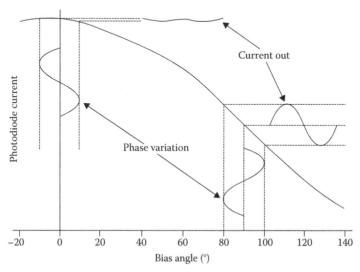

FIGURE 4.3
Sensitivity at 0° and 90° bias.

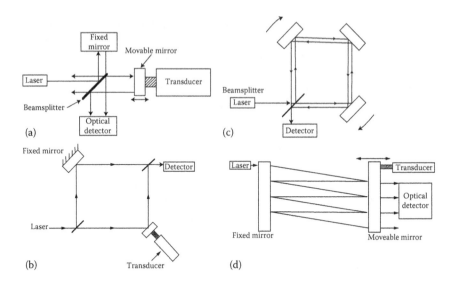

FIGURE 4.4
(a) Michelson interferometer, (b) Mach–Zehnder interferometer, (c) SAGNAC interferometer, and (d) Fabry–Perot interferometer. (After Krohn, D.A., *Fiber Optic Sensors: Fundamentals and Applications*, Instrument Society of America, Research Triangle Park, NC, 1988.)

detector. The sensing path is from the laser, through the beam splitter, off the transducer/mirror, and off the beam splitter to the detector. The Mach–Zehnder configuration includes a fixed and moving mirror/transducer, with clockwise and counterclockwise propagation paths. In the SAGNAC interferometer, the clockwise and counterclockwise paths produce a phase difference upon rotation of the entire system.

Finally, in the Fabry–Perot interferometer, the transducer/mirror is partially transmitting, so that a change in the total path or length or total number of reflections at the second mirror produces a change in the total light intensity at the detector.

The four common fiber interferometer configurations are illustrated in Figure 4.5. Here, the free-space optical paths are replaced by fiber-optic waveguides, the mirrors are multilayer dielectric coatings, the couplers are fused biconical taper couplers or integrated optical couplers, and the lasers and detectors are coupled to the input and output fibers, respectively.

4.4 Fiber-Optic Magnetometer

The fiber-optic magnetometer is representative of the class of interferometric sensors. A magnetometer design is discussed to illustrate many of the

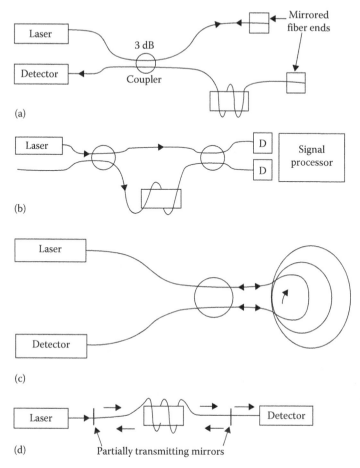

FIGURE 4.5
(a) Michelson, (b) Mach–Zehnder, (c) SAGNAC, and (d) Fabry–Perot interferometers.

design considerations applicable to the entire class of sensors. The fiber magnetometer is implemented as an all-fiber Mach–Zehnder interferometer in which one branch has been made sensitive to magnetic fields by coupling to a magnetostrictive material. The Mach–Zehnder magnetometer is capable of sensing very small axial strains in the sensing fiber.

An early potential application of the fiber magnetometer was a sensor in mine applications. This utilized the improved sensitivity over flux gate magnetometers in proper detection of perturbations of the local magnetic field due to a moving vehicle containing large amounts of ferromagnetic material. Potential applications of the fiber magnetometer include [6] tank, truck, aircraft, and ship detection; electric current monitor; geophysical magnetic disturbance monitor; magnetic compass; metal detector; VLF receiver; lightning monitor; traffic controller; and multi-influence target detector.

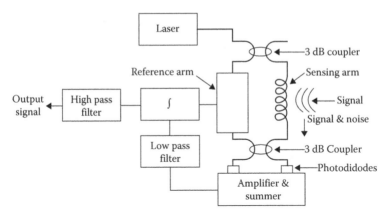

FIGURE 4.6
Mach–Zehnder fiber-optic interferometer employing phase-locked homodyne detection.

Figure 4.6 illustrates the configuration consisting of a laser source, 3 dB beam splitter, two optical fiber paths, another 3 dB beam combiner, and a photodiode detector. One of the fiber paths is designed as the sensing fiber and the other as the reference fiber. If the path lengths are identical so that the two light waves arrive at the combiner exactly in phase, the detector output will be a maximum. If one path length is an odd number of optical half-wavelengths longer or shorter than the other, the two beams will interfere destructively at the combiner and the detector output will be minimized.

If one fiber is elongated with respect to the other, the detector output will go through one complete cycle each time the difference in optical path lengths increases by one wavelength. Because the optical wavelength is very small and the fiber can be made very long in terms of wavelength, the device can become an extremely sensitive detector of any phenomenon producing a change in fiber length. Many magnetic materials are magnetostrictive, undergoing dimensional changes upon magnetization. Magnetostrictive materials have been coupled to the sensing fibers of Mach–Zehnder interferometers to form sensitive magnetometers [1].

The fiber magnetometer sensing fiber can be coupled to the magnetostrictive material by coating the bare fiber with the metal or by bonding the fiber to bulk shapes of material, as illustrated in Figure 4.7. The bounding method shown in Figure 4.8 is most commonly employed because it yields the largest system sensitivity and is most suitable for mass production.

4.5 Fiber Acoustic/Pressure Sensors

Fiber sensor transducers can be designed to sense a wide variety of physical phenomena. One of the more common examples is the use of an acoustic

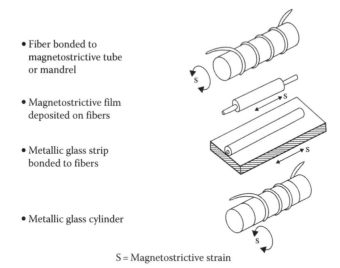

- Fiber bonded to magnetostrictive tube or mandrel

- Magnetostrictive film deposited on fibers

- Metallic glass strip bonded to fibers

- Metallic glass cylinder

S = Magnetostrictive strain

FIGURE 4.7
Ways of coupling fibers to magnetostrictive material.

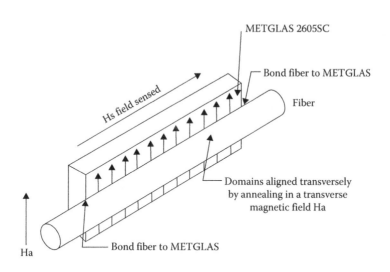

METGLAS 2605SC

Bond fiber to METGLAS

Fiber

Hs field sensed

Domains aligned transversely by annealing in a transverse magnetic field Ha

Bond fiber to METGLAS

Ha

FIGURE 4.8
Coupling fibers to METGLAS 26055SC.

pressure transduction mechanism in an acoustic and pressure fiber sensor. Such a transducer is illustrated in Figure 4.9. System applications of these sensors include detection of ships, submarines, sea state indicators, pressure gradient measurement, and multi-influence target detection. Section 4.18 contains a fiber-optic system design example for a fiber-optic sonar dome pressure transducer.

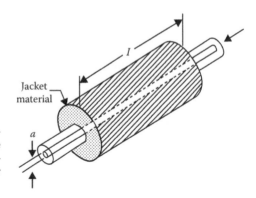

FIGURE 4.9
Acoustic pressure transduction mechanism. Increase in ambient pressure causes jacket to elongate in the *I* direction. This strain is transferred to the fiber increasing its optical phase length.

4.6 Optical Fiber Characteristics

Single-mode fibers used in fiber sensors place stringent connectivity tolerances on the system relative to those in the multimode fiber system. Rayleigh scattering caused by the glass imperfections is the intrinsic optical loss mechanisms in the fiber. This loss generally decreases with increasing wavelength from the visible to the infrared. Because of this and other attenuation factors, the best region for transmitting signals through optical fibers is between 1100 and 1600 nm. The region from 800 to 900 nm is also relatively favorable for transmission; since sources and detectors for this wavelength range are better developed, it is often the operating region for fiber sensors.

The analysis of single-mode fibers is fairly straightforward, since waveguide theory produces approximations to the mathematical description of waveguide parameters. Coupled with knowledge of the core cladding dispersion characteristics, this permits the determination of the range-bandwidth product. With single-mode fiber transmission, an operating wavelength can be determined where both attenuation and dispersion factors are at optimal values for the selected fiber. Of more direct importance to the fiber system is a consideration of the various sources of induced phase modulation and the fiber system response to these perturbations. Variations of the normalized propagation constant due to variations of core diameter and refractive index have been shown to be negligible [7]. This leaves the consideration of various environmental effects such as temperature and pressure (or acoustic) sensitivity of the fiber length. If the optical path lengths of the two sensor arms are equal to within the coherence length of the laser, the desired interference effect occurs. A change in the relative phase of the light between fibers appears as a change in fringe position or light intensity at the detector.

A detailed analysis of the effects of strain induced by the difference in the thermal expansion coefficients between the various regions of the sensor is necessary for the prediction of thermal sensitivity. Temperature effects

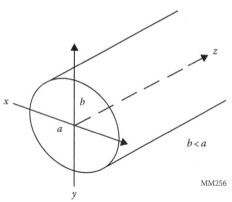

Modal birefringence $B = (\beta_x - \beta_y) \lambda_o/2\pi$

Beat length $L = \lambda/B$

FIGURE 4.10
Optical fiber with elliptical cross section.

occur as well, particularly with increasing thickness of the jacket material. Sensitivity to static pressure changes is low compared to other effects; however, longitudinal strains due to acoustic signals can provide a pronounced perturbation of the system signal. Stabilization techniques for these effects are described in a later section.

Another fiber consideration arises from the fact that the coherent detection process used in the fiber interferometer requires that the polarization at the output of both fibers remains the same. This can be solved most efficiently with polarization-maintaining fibers. When the fiber is wound on a drum or mandrel, linear polarization-maintaining elliptical fibers as illustrated in Figure 4.10 are preferable to the circular polarization-maintaining fibers.

4.7 Fiber Transducer Considerations

Transducer design for the fiber magnetometer is again representative of many necessary considerations for other fiber sensor systems, such as fiber pressure and acoustic sensors. As mentioned previously, two approaches can be considered in the coupling between the magnetic material and the fiber. A coating of magnetic material can be applied directly on the fiber, or the fiber can be attached by an adhesive to an already formed heat-treated magnetic strip or film. Each approach requires a unique set of processing considerations.

A number of coating techniques are available. These include vapor deposition, sputtering, electrodeposition, and electroless deposition. Each has been used successfully in the coating of recording discs for information storage [8]. The application of magnetostrictive active magnetic coating to an optical fiber is a difficult process. Both sputtering and vapor deposition are very directional and may not produce a uniformly coated fiber. Sputtering deposition rates are slow (less than 1000Å min^{-1}), compared to vapor deposition (up to $1 \mu\text{m min}^{-1}$). Magnetostrictive coatings approaching $40 \mu\text{m}$ are necessary in the magnetometer design. The problem of directionality is eliminated with electrodeposition or electroless deposition. The former requires a thin electrical conducting layer deposition on the fiber prior to the process. This approach has been used in the successful electrodeposition of nickel and iron alloys on fibers; however, electrodeposition of binary and multicomponent alloys becomes difficult because of the different deposition potentials of the individual components. Thus, electroplated alloy compositions do not readily correlate with solution compositions. These problems can be overcome by a change of solution type or adjustment of solution composition.

Electroless deposition has an advantage over electrodeposition in that a thin conducting layer is not necessary beforehand. This process requires a suitable metal-bearing salt, a reducing agent, and an appropriate catalyst. It has been used successfully in the deposition of cobalt for metallic film disks. Problems with alloy compositions are similar if not more severe than with electrodeposition. For optimum magnetic response, the coated fiber must be heat treated. Heat treatment relieves deposition strains, homogenizes the coating, and adjusts the ferromagnetic domain structure. The goal is to obtain the maximum possible value of the quantity $d\lambda/dH$ for the coated fiber. Here $d\lambda$ is the change in length of the coated fiber (per unit length) due to magnetostriction, and dH is the change in the field applied along the length of the fiber. This rate of change of length with field depends in a complex way on both structure-insensitive factors such as magnetic material, the anisotropy constant, and the level of magnetization, and structure-sensitive factors, including the ferromagnetic domain configuration and orientation, grain size and orientation, residual stress, inclusions, and the sample demagnetization factor. We note here that the maximum $d\lambda/dH$ value for a given sample need not necessarily coincide with the maximum dB/dH value, where B is the magnetic flux density.

In coated fibers, the optimum value of $d\lambda/dH$ is unlikely to be obtained because of the limitations imposed on the structure-intensive parameters by the material systems that can be readily coated from solution and on the structure-sensitive properties by the limited processing and heat treatment that can be imposed. For example, the thin layer of copper used to provide electrical conductivity in the electroplating of iron–nickel alloys will diffuse under heat treatment at 1000°C, resulting in degraded magnetostrictive properties [9]. A longitudinal annealing field could be applied with a solenoid, but the only way to apply a circumferential field would be to pass a DC current along the wire.

Stresses can also develop in the fiber either from any bending that might be necessary in the configuration of the magnetic-coated-fiber part of the sensor or from differential thermal contraction between the coating and the optical fiber upon cooling from an annealing treatment. The bending stresses should not be unique to the coated fiber provided the sensor configuration of each unit (coated versus attached fiber) is the same. Similarly, the bending stresses are only related to the radius of curvature of the bend and will not differ for the two designs. They are, however, unidirectional and nonuniform, and will contribute to a birefringence through the piezooptical coefficients.

Differential cooling stresses following heat treatment result in a compressive stress along the fiber as well as a compressive loop stress. Such stresses result in a change in the optical indices of the fiber and thus a phase change in the light transmitted compared to a similar length of unstressed fiber. The alternative approach of fixing the fiber to an already processed and treated magnetic material deals with the attachment problem. The adhesive acts as a conduit for the magnetostrictive strain from the magnetic material to the fiber. As such, its elastic strength and viscoelastic behavior must be suitable to obtain fiber strain as close as possible to that in the magnetic material. This is similar to the problem dealt with in the fabrication of strain gauges. The opportunity to select the optimum magnetic material and to use the optimum heat treatment makes the last approach discussed preferable in most designs. The optimum achieved is an appropriate balance between the sensitivity to field variation (i.e., $d\lambda/dH_{app}$), the bias field applied, and hysteresis effects.

The amorphous alloy METGLAS 2605SC (an Allied Corporation product) has been shown to exhibit large magnetoelastic coupling. Sensors fabricated by bonding single-mode unjacketed optical fibers to strips of METGLAS and to cylinders formed by coiling strips of the material have demonstrated sensitivities greater than those obtained with the plated fiber sensors. It has also been reported that annealing the METGLAS in the presence of a magnetic field in the direction of the strip width increases the sensitivity and greatly reduces Barkhausen noise produced by domain wall movement [10]. Depending on the fabrication procedure, unbalanced radial stresses induced in the fibers may degrade performance due to polarization shifts caused by strain-induced birefringence. This effect is minimized by sandwiching the fiber between layers of METGLAS ribbon.

Finally, the material used to bond the fiber to the METGLAS must couple the magnetostrictive strain to the fiber without introducing spurious signals due to thermal expansion or ambient pressure changes. It must not relax excessively with time or temperature. Care must be exercised in the sensor design to avoid resonances and to preserve the frequency response over about a 5000 Hz bandwidth.

4.8 Fiber Sensor Laser Selection

Use of a semiconductor laser is essential to the ultimate miniaturization of the fiber sensor system. The advantage of high efficiency, narrow spectral width, low power consumption, high coupling efficiency (modest output beam divergence), and optimum wavelength matching at single longitudinal mode operation available in commercially available devices preclude the use of HeNe or other types of laser sources in many applications. Although a wide variety of laser diodes is available, the group consisting of $Al_xGa_{1-x}As$ $(0 < x < 0.5)$ in a double-heterostructure configuration is the most suitable for fiber sensor design. Low-threshold continuous operation at room temperature has been achieved, and a stable Gaussian beam is produced without the use of a spatial filter or other apodizer. Feedback diodes are typically incorporated in the devices to maintain power levels. Proper heat sinking diode junction thermistor temperature controls have been developed to control variations of emitted wavelength with temperature. Most laser diode degradation mechanisms have been identified and eliminated so that 10^6 h lifetimes are projected. Finally, the output spectrum of these devices shows no significant shift with current level changes up to 1.5 times threshold.

Wavelengths between 630 and 900 nm, spanning the fiber low-loss region, can be achieved with appropriate Al concentrations. Injection laser beam divergence is asymmetrical, with typical half angles of less than $5° \times 10°$–$10° \times 30°$. The major portion of the output of these lasers falls within an ellipse of $f/6 \times f/12$–$f/1.8 \times f/6$ (numerical aperture values of 0.17×0.08–0.55×0.17). Although circularizing optics in the form of cylindrical lenses or prism combinations can render this input circular, and spherical lenses then used to couple the laser output to a single-mode fiber, this arrangement requires extensive lens positioning, fixturing, and unnecessarily large volume. A preferred approach for coupling the laser output is to pigtail the single-mode fiber directly to the laser diode package. This arrangement provides excellent coupling efficiency at a reduced volume and is available with several of the commercial devices. Although the coupling coefficient between the laser and the fiber depends on the far-field pattern of the laser and the diameter and acceptance angle of the fiber, coupling loss for a well-designed system is typically -2 to -3 dB.

Extremely low current thresholds are available (less than 12 mA) and in fact the electrical-to-optical power conversion efficiency of injection lasers is among the highest of all candidate lasers (up to 0.45 W/A for diodes in the lasing region). The diode emits a linearly polarized TE mode with the electric field vector parallel to the p–n junction plane. Typical coherence length for a single longitudinal mode laser is about 6 m for a 50 MHz linewidth (4 km for 75 kHz). Noise performance of these diodes is excellent. Most of the output is generated via a quantum statistical process by the spontaneous emission of light and by small fluctuations in the pumping current and temperature variations.

Temperature affects all of the material parameters of a semiconductor that are important for lasing. These factors combine to give an overall slight decrease in output as temperature rises. Many laser diode modules contain a small monitor photodiode that is used in a feedback loop to control the drive current in order to maintain the laser output at a constant level. To the extent thermal or noise factors play a significant role on fiber sensor performance, these compensation schemes can be easily employed.

A final consideration in the choice and operation of the laser diode is optical feedback. Various optical discontinuities in the fiver sensor transmission system can reflect light back to the source and deteriorate its operating characteristics. If such optical feedback becomes excessive, performance of the sensor system may be impaired. In this case, an optical isolator would be required between the laser pigtail fiber and the system fiber connector. The isolator rotates the plane of polarization of the feedback light by $\pi/2$ so that it can be filtered by a polarizer.

4.9 Laser Frequency Stability Considerations

The major difficulty in implementing the interferometric fiber sensor involves the requirement for a coherent and stable light source. Accuracy of sensing is affected directly by the laser source and any mode jumping, phase noise, or mode partition noise. Laser structures with the most stable operations possible are chosen for the fiber sensor system. For applications in which size and power consumption are not limited, a gas laser should be considered since the gain curve of the total linewidth is typically on the order of 1 GHz [11]. With the gas laser, it is relatively easy to define the wavelength to parts per million; in some cases, the HeNe laser can be locked to the center-gain curve with stability on the order of parts in 100 million [12]. This parameter is very important because changes in the operating frequency result in time delays due to fiber dispersion, even in the low-dispersion operating regions of the single-mode fiber.

Along with the effect of dispersion and frequency instability, the performance of the sensing system can be affected by the presence of phase noise. Phase noise occurs due to imperfections in the laser cavity, laser drive circuit noise, and resonance within the laser structure itself [11]. Some control of excess phase noise is possible through implementation of a feedback control system in the laser drive. Tight control of the laser drive current is necessary because the laser resonance is a function of the drive current. Typically, a change in drive current of 1 mA will result in a change of several gigahertz in lasing frequency. Other enhancements such as modification to the laser emission linewidth through the use of an external cavity and the use of thermoelectric cooling of the laser body provide only minimum improvements. Laser source stability improvements are necessary for full implementation of the interferometric fiber sensor. Further developments are also necessary before laser manufacturers offer the optimized laser structured in a wide variety of operating frequencies; therefore, external stabilization techniques must be employed as an interim solution to the stabilization problem. One such approach to achieving stable operation of the laser source involves the use of an external electronic feedback circuit and the fiber-optic ring resonator.

Experimental measurements of the output spectrum produced by single-mode solid-state semiconductor lasers [13] have resulted in observations of instabilities, including mode jumping, phase noise, mode partition noise, line broadening, and frequency shifting. Unstable operation in fiber-optic interferometric sensing systems has also been observed. Although stability problems exist for certain applications, semiconductor lasers have been demonstrated with excellent performance in communications and coherent transmission under carefully controlled operating conditions (for a complete discussion, see Ref. [14]).

In communications, the signal bandwidth can and usually does exceed the laser linewidth [12]. For example, a communications system using a directly modulated laser can support a signal bandwidth of 2–3 GHz. This bandwidth is much wider than the laser linewidth (a laser linewidth of 10 kHz was reported through the use of a fiber-optic ring resonator). Because of this wide bandwidth, some of the problems associated with the frequency instability are not critical to operation in noncoherent communication systems. In sensors and interferometers, however, the signal bandwidth may be very small in comparison to the laser linewidth. For example, if the sensor is designed to monitor ambient room temperature or perhaps the magnetic profile of a slowly moving object, the signal sensed may be in the near-dc range, while the linewidth of the laser used in the sensor itself may be in the 1 GHz range. If the laser linewidth suddenly broadens or changes frequency during operation, a resulting amplitude or phase fluctuation may be produced by the sensor, thus degrading performance due to "random noise." The need for frequency stability and low noise in the light source for sensor applications is extremely important and may even impose more stringent requirements on laser operation than can be presently achieved in many commercially available devices.

Some laser structures, such as short cavity, external cavity, distributed feedback (DFB), and rare-earth-doped heterostructure, offer various degrees of stability; but in general, the ideal stabilized laser diode remains to be developed. Various stabilization techniques, such as thermoelectric control of the laser structure, feedback control of the laser injection current, the use of the external fiber-optic ring resonator, and so on, are designed to improve laser stability [14].

4.10 Couplers and Connectors for Fiber Sensors

The fiber-optic single-mode coupler is an essential component of the fiber sensor system, serving as the power divider and combiner in the interferometer.

Single-mode connectors are utilized for optimum performance and alignment tolerances in connectorization of fiber to fiber. Any alternative bulk optics approach employing beam splitter–lens combinations to split and recombine optical powers in the interferometer will surely require excessive space and alignment tolerance ruggedization.

The guided-wave equivalent of the bulk optic beam splitter is the two-fiber four-port coupler. These couplers are based on the fact that when two fiber cores are brought sufficiently close to one other, their modes become coupled through the evanescent fields. This results in a reciprocal power transfer from one fiber to the other, with the power transfer ratio depending on the core spacing and interaction length. A wide range of fabrication procedures have been developed; however, one technique appears to be superior [15]. This process employs the fused biconical taper technology, in which two fibers are twisted around one other, flame heated, and fused, while the two stages of the fusion station are moved apart. The biconical taper brings the two fiber cores close to one other but simultaneously decreases the core diameters considerably, which increases the mode spot radius, both of which contribute to strong coupling through the evanescent tails of the guided modes. This technique provides a very rugged power divider without fiber material removal. Several forms of commercial devices are available.

Fiber-to-fiber connections can be either permanent or semipermanent joints achieved by bonding or thermal fusion, or detachable joints using demountable connectors. The first type provides higher reliability and stability under field conditions; but from the viewpoint of component interchangeability and testing during system development, it is more reasonable to use demountable connectors for coupling the optical source pigtail to the splitter, and the splitter and combiner to the sensing and reference fibers.

Many conventional connector manufacturers offer unique fiber-optic designs. Because of the small core diameter of single-mode fiber, the positioning accuracy of the connector becomes a challenge, in that the attainment of a coupling loss smaller than 1 dB requires the fiber end-to-end separation be maintained below about 30 μm, the core-to-core lateral offset below 1–1.5 μm, and the axis-to-axis tilt angle below 1° (figures are for a 1300 nm operating wavelength and will be slightly less stringent at 850 nm) [16]. The general principle of the field assembly connector is to use precision-machined ceramic ferrules, with the fiber inserted and epoxied into the central hole, and the ferrules then assembled into a coupling nut. Additional active fiber-positioning techniques can be employed to provide further alignment optimization. A final consideration in the coupling of single-mode polarization-preserving fibers is the potential polarization coupling between one of the two fundamental HE mode polarization states due to alignment error between the main polarization planes of the two joined fibers. This effect is proportional to $\sin^2 \theta$, where θ is the rotation angle [17]. This coupling can be minimized with proper design.

As a final consideration we note that since the area of detectors is usually larger than the fiber core area, it is conceptually a straightforward matter to couple light from the fiber to the detector. Coupling loss arises mostly from Fresnel reflections and can be minimized by refractive index matching or by the use of an antireflection layer on the detector.

4.11 Fiber Sensor Detector Considerations

There exist a large number of different photodetectors, ranging from the large photoemissive and photomultiplier vacuum tubes to solid-state and semiconductor devices. The latter are the appropriate choice for fiber systems, with the most practical operating in the near infrared range. These are the silicon p–n and PIN diodes operated in reverse-biased photoconductive mode. These devices provide a change in conductivity proportional to the optical generation rate. When operated in the third quadrant of the IV characteristic, current is essentially independent of voltage but proportional to the generation rate of electron–hole pairs. Such a device provides a useful means of measuring illumination levels or of converting time-varying optical signals into electrical signals.

When an election–hole pair is created in the depletion region, the electric field sweeps the electron to the n side and the hole to the p side. It is therefore desirable to make the depletion width W large enough so that most of the photons are absorbed within W (depletion-layer photodiode) rather than in the neutral p and n regions. The appropriate W is chosen as a compromise between sensitivity and speed of response, since W must not be so wide that the time required for drift of photogenerated carriers out of the depletion region is excessive.

Internal gain is produced by applying external voltage sufficient to extend the high-field region into the two regions adjoining the junction. Accelerated carriers now have an increased probability of colliding with the semiconductor atoms, freeing even more carriers. The PIN structure includes an intrinsic region tailored to meet particular drift and multiplication effects. For detection of very low optical signals, the photodiode is often operated in the avalanche region (avalanche photodiode). In this mode, each photogenerated carrier results in a significant change in the current due to avalanche multiplication. Significant current gain can be achieved, but at the expense of the introduction of excess noise. The APD is a more expensive component than the PD, and requires a greater complexity for the bias circuit and power supply.

Numerous commercial PIN diodes are available for operation at around 850 nm. This is, in fact, near the peak sensitivity of silicon devices, with a typical range up to 400–600 μA mW^{-1}. Minimum detectable signal levels are on the order of 5×10^{-15} W. These devices have a wide dynamic range and good linearity properties. Inexpensive units have typical operational biases from 2 to 100 V, with active areas from a few millimeters to 1 cm. They generally have uniform responsivity over the entire active area, and are quite stable with time and temperature.

The PINs are typically packaged in modified versions of standard TO-5 and TO-18 cans, with either flat glass or convex lenses. Several special package designs are available for fiber-optic applications where the detector is mounted in a connector shell with a pigtailed fiber. Since the detector active area is not typically well matched with the size of the fiber, coupling losses as well as volume requirements can be reduced significantly through the use of a specially packaged detector, so that such an approach is well suited for fiber sensor systems.

4.12 Fiber Magnetometer Applications

To sense a vehicle with a fiber magnetometer, the vehicle must be the source of a magnetic field. Such a field may exist or may develop through the interactions of the vehicle's ferromagnetic material with the earth's magnetic field. The vehicle-generated field is usually small or nonexistent. For the latter field, there is direct correlation between the local earth's field strength and direction and that of the field generated. The total local field is then the vector sum of the two fields and can be considered to be the perturbed earth's field.

The magnetic sensor in the near vicinity must adequately detect the perturbation above the background earth's field. Adequate detection leads to the determination of both vehicle type and location. The model of a vehicle in the earth's field is that of a magnetic moment, the magnitude of which is determined by the vehicle volume and geometry as well as the earth's field. The moment direction can be expected to follow closely the field direction of the earth, although the vehicle geometry may dictate that they are not exactly parallel. The field of a dipole moment is well known and can be expressed in polar coordinate form in terms of R and θ, where R is the distance or range from the dipole moment and θ is the angle between the R direction and the moment direction. Two independent measurements are required to evaluate either R or θ with additional information required to evaluate vehicle type. If two directional sensors are arranged orthogonally, R and θ can be determined. The easiest situation to analyze is a common plane formed between the sensor direction and the vehicle moment. If this is not so, only the field component for the moment in the plane of the sensors is detected. Knowledge of the moment angle with the sensor plane is necessary if evaluation of R and θ is to be made. This can be done if a third noncoplanar sensor is used or if the angle can be estimated from the direction of the earth's field. Detailed information on R and θ is necessary if, for example, a mine is to be activated at the most appropriate time.

The analysis above assumes knowledge of the vehicle type and its movement. Only the low-frequency magnitude of the perturbation signal was considered. To obtain information on vehicle type, the noise spectrum in the signal can be used and correlated with expected noise for a specific vehicle type. This leads to the appropriate value of the moment, and requires a transfer of field-time measurements into the frequency domain.

4.13 Fiber Sensor Operation

The basic fiber sensor output voltage will pass through a series of maxima and minima as the local magnetic field intensity increases or decreases, provided the magnetic variations are great enough to change the effective fiber length by many optical wavelengths. The number of output peaks occurring in any time interval is a measure of the total change in field strength during that interval. The result will, however, be the same for increasing and decreasing fields, and the negative and positive changes within any measurement interval will add on the basis of absolute magnitude irrespective of algebraic sign (e.g., suppose that the field went from 0.5 to 1.0 Oe and back to 0.5 Oe within an observation interval). The net change in field intensity is zero, but the peak counting technique will indicate a net change of 1.0 Oe. Furthermore, unavoidable drift in differential optical phase lengths between the reference sensor fibers will produce spurious maxima and minima that cannot be distinguished from desired signals.

The drift and ambiguity problems can be overcome by incorporating a phase-locked feedback system. Such an arrangement is show in Figure 4.11.

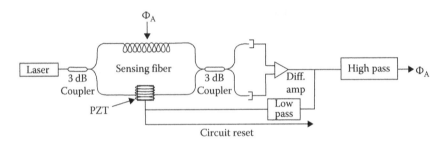

FIGURE 4.11
Active homodyne detection.

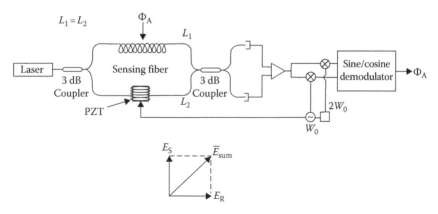

FIGURE 4.12
Homodyne using phase-generated carrier (carrier generated by PZT).

Here an electrically controlled optical phase shifter has been introduced into the reference fiber path. The detector output voltage is amplified and passed through a low-pass filter to provide the control voltage to the optical phase modulator. Figure 4.12 shows the optical power from the summing device when the reference arm and the sensing arm outputs are in phase quadrature. Assuming the output magnitudes are equal, the vector sum is $\sqrt{2}$ times the amplitude. If a voltage is applied to the phase modulator in a direction to increase the phase difference, the magnitude of the sum will diminish. If this voltage is the amplified output of the photodetector, the reference and sense arm phase difference will be maintained at exactly 90°. AC voltages appearing at the detector output that are above the cutoff frequency of the low-pass filter will not reach the phase modulator. These voltages are then representative of signal disturbances above the filter cutoff frequency. If all unwanted drifts and instabilities occur below the filter cutoff and the loop gain is high, the sensor is stabilized against drift and the two arms are locked in quadrature phase.

Consider, for example, the scenario illustrated in Figure 4.13, in which a vehicle follows a path along a chord subtended by a 90° angle at the center of the 120-m-diameter circle describing the sensitivity limit. It will be approaching the sensor for 42.4 m and receding for a like distance. If it is traveling at 7.3 ft s⁻¹, it will require 19 s to reach the closest point. If we take the lowest-frequency response required as 1/19, we arrive at a cutoff frequency of about 0.05 Hz. Drift mechanisms must be controlled to operate below this frequency (e.g., by insulating the sensor from thermal and pressure changes). If the sensor response extends from a very low frequency to 5000 Hz, it encompasses commercial power-line frequencies. Frequencies in the range of 50–60 Hz can be rejected with notch filters.

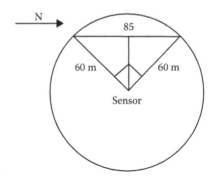

FIGURE 4.13
Moving vehicle target.

4.14 Fiber Sensor Signal Processing

The fiber sensor is analogous to the classical phase-locked-loop (PLL) circuit used in a phase demodulator application. As a target's signature influences the sensor, it causes a phase shift in one of two coherent signals applied to a phase comparator network. A voltage proportional to the induced phase shift is generated and applied to an internal phase shifter on the reference leg. The signal is then high-pass filtered to eliminate external noise-induced bias below around 0.05 Hz in a detection loop. This action forces phase tracking between the two signals; as one signal's phase is modulated by the target, the reference signal's phase is controlled to track the effect, just as in a conventional PLL. The output of the system is the control voltage driving the internal phase shifter, which directly represents the phase shifts due to the target's influence on the sensor transducer; thus, the phase shifts are demodulated. Identifying the classical blocks in a PLL, the combining directional coupler and photodiode represent a phase detector, the signal fiber produces the phase-modulated signal, and the piezoelectric phase shifter and reference fiber are analogous to a voltage-controlled oscillator (VCO). It is worth noting that the phase modulator is not precisely identical to a VCO in that the loop filter cannot induce a change in the frequency of the source. This difference becomes significant since a frequency drift of the source remains uncorrected and manifests itself as a loop bias (i.e., is tracked). Loop filter transfer functions are selected to provide the desired steady-state and transient responses (phase domain only) in accordance with the classical PLL demodulator.

4.14.1 Reference Phase Modulation

A phase modulator is required in the reference arm to maintain phase lock at the quadrature condition and to correct for thermal drift. Such a modulator is formed by winding a portion of the reference arm fiber around a piezoelectric cylinder. Dimensional (diameter) changes in the cylinder produce axial strains in the fiber, changing the path length and total phase shift. These devices have the desirable properties of requiring very little power for low-bandwidth (slow) phase changes and of operating in a voltage rather than a current mode. The latter produce magnetic disturbances that could lead to instabilities in a sensitive fiber magnetometer.

4.14.2 Fiber Sensor System Noise

Both inputs to the phase detector will have a very high signal-to-noise ratio (SNR); since the sensor is a phase sensor, it is operated with the carrier power well above the threshold level where a loop can phase lock. Because of ample signal power at both inputs of the phase detector, the phase stability of the laser and the self-noise of the photodiode and loop amplifiers will be the SNR-limiting factors in the sensor.

The laser phase noise is applied to both inputs of the phase detector with slightly different time delays. If the phase fluctuation power spectral density is known, the effects of this phase noise on the detector output can be predicted as follows. The phase detector output $\beta(t)$ is proportional to the phase difference between the two signal paths:

$$\beta(t) = A \cos(\omega_c t + a_1) + A \cos(\omega_c t + a_2). \tag{4.1}$$

We are interested in the noise component of the detector output $E(t)$. The photodiode produces an output proportional to the intensity of $\beta(t)$, which is the sum of two phasors:

$$|\beta(t)|^2 = 4A^2 \cos^2\left[\frac{(a_1 - a_2)}{2}\right]$$

where a_1 and a_2 are the phases of the input signals. Both phases have a common noise component $\varphi n(t)$ but delayed differently in time:

$$a_1(t) = \delta(t) + \varphi_n(t) \text{ and } a_2(t) = \delta(t) + \varphi_n$$

(If $\tau = 0$, the source phase noises will completely cancel.) The allowable delay difference between the two paths is obtained by studying the autocorrelation function of $\varphi_n(t)$; high correlation coefficients result in a high degree of phase noise cancellation.

The time autocorrelation function $R\varphi(t)$ is (assuming an ergodic process)

$$R\varphi(t) = \lim_{T \to \infty} \frac{1}{2T} \int_{-T}^{T} \varphi_n(t)\varphi n(t - \tau) \tag{4.2}$$

$$= \frac{1}{2T} \int\limits_{-\infty}^{\infty} S_\phi(\omega) e^{f\omega t} d\omega$$

where $S_\phi(\omega)$ is the two-sided power spectral density of the source phase fluctuations $\phi_n(t)$. Once $S_\phi(\omega)$ is known, an allowable system limit on τ, the differential time delay between the two signal paths, is set.

Even if the source is relatively noisy, its effects can be minimized by minimizing the absolute phase differences between the two fiber paths. The detector and loop amplifier noise are often the dominant system noise sources. In addition to the random zero-mean fluctuations of instantaneous phase in the laser, the system will have two major bias components to contend with: temperature differentials between the two signal fibers and frequency drifts in the laser. Both biases can cause phase biases that the loop attempts to eliminate via its tracking mechanism. As bias signals become large, the piezoelectric phase shifter is not able to track them since it cannot cause phase shifts of many radians. (Note that in a classical PLL, the typical VCO can shift many hundreds of megaradians when required.) When the phase shifter is saturated, further bias input causes the loop to experience a transient of 2π radians in phase state, with the frequency state being a variable outside loop. At this point, the loop will be nearly saturated with opposite polarity if its total dynamic range is 2π radians; thus when the loop resets its steady state automatically, it will return to the middle of the signal range. A loop dynamic range of 4π radians minimizes the amount of time the loop is in saturation without creation of undue performance requirements on the phase shifter.

The differential phase shift $\Delta\phi$ due to frequency shift Δf is $\Delta\phi(2\pi\Delta1)/v_0\Delta f$, where v_0 is the velocity of propagation in the fibers, and Δl is the physical path length differential. The differential phase shift due to temperature change from steady state of the two fibers is

$$\Delta\phi = \left(\frac{2\pi}{\lambda}\right)(TC_1 l_1 \Delta T_1 - TC_2 l_2 \Delta T_2) \tag{4.3}$$

where
  TC_1 and TC_2 are the temperature coefficients of expansion in the two signal paths
  l_1 and l_2 are the physical lengths of the two signal fibers
  T_1 and T_2 are the differences in the two signal paths' ambient temperatures from a reference temperature

A temperature-related phase bias is introduced into the loop unless both signal paths have identical temperature coefficients of expansion and physical lengths and are at the same ambient temperature.

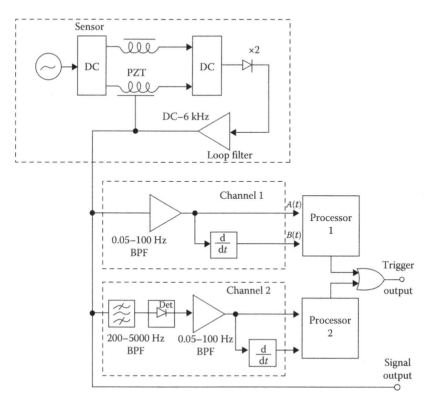

FIGURE 4.14
Signal processing block diagram.

A common fiber sensor signal-processing approach is illustrated in Figure 4.14. The sensor output is split into two bandwidths, one that cover target dynamics (0.5–100 Hz in channel 1) and one that is sensitive only to high-frequency signals (channel 2). The output of high-frequency bandpass filter is detected and amplified. The envelope amplitude of the high-frequency signals behaves like a target with a large steady-state signal; thus, the envelope of the channel 2 signal is processed in the same way as a channel 1 signal. Figure 4.15 illustrates the basic output of each channel—a "smoothed" output and its derivative. At time t_0 in the figure, the target is at its point of closest approach to the sensor, and the processor generates a trigger. The time t_0 is detected by requiring that the signal derivative $B(t)$ change while $A(t)$ exceeds a threshold V_{th} as shown. If either processor 1 or processor 2 detects this condition, a trigger is generated. The processors are realized using very simple analog circuits. One final alternative to this approach is the use of a carrier generated by a frequency-modulated laser. This detection scheme is illustrated in Figure 4.16.

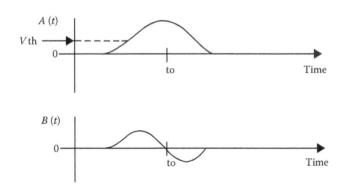

FIGURE 4.15
Output signals.

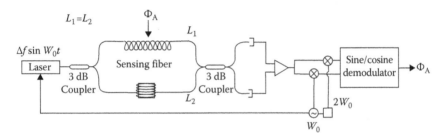

FIGURE 4.16
Homodyne using phase-generated carrier (carrier generated by frequency-modulation laser).

4.15 Environmental Stabilization

Fiber sensors are extremely sensitive mechanical strain detectors, responding to any strain-producing influence in either fiber path. Optical fibers exhibit a change in phase length in response to ambient pressure changes. The magnitude of the effect is strongly influenced by the protective jacket material on the fiber. Rubber or rubberlike elastomeric materials cause the greatest pressure sensitivity. Coatings are selected to minimize pressure (static and acoustic) sensitivity and to balance the sensing and reference paths.

Thermal expansion changes the optical length of the fibers. The combined thermal expansion effects of the fiber, its jacketing, the magnetostrictive or other transducer material, and the mechanical support material must be considered. Some control over the balance between the reference and sensing paths is afforded by the material selected to support the reference fiber.

The laser frequency and intensity are affected to some extent by temperature and power supply variations. Laser frequency changes produce differential phase changes and hence noise if the two optical path lengths are not exactly balanced (i.e., there will be a phase shift proportional to the frequency difference multiplied by the difference in path lengths). This effect is near the system noise levels for typical fiber sensors.

Fiber sensor models present to the detector the sum of two sinusoids of equal and constant amplitude but of variable differential phase so that the resultant intensity is a measure of the differential phase. Light-intensity modulation by the laser or fibers produces apparent phase noise and must be minimized. A feedback-stabilized laser is typically employed. Procedures used to reduce unwanted phase modulation in the fibers reduce intensity modulation as well.

Optical standing waves can exist between two reflective mismatches, creating, in effect, a Fabry–Perot resonator. The transmission of such a resonator depends on its length at any particular frequency. Possible sources of reflections are separation of fiber ends at connectors and/or misalignment at connects, couplers, or splices. Temperature variations may change the separation of reflective mismatches and alter the transmission loss. Careful attention to the assembly of components is required to minimize transmission losses, to lower the Q of the resonant sections, and to reduce frequency/thermal transmission variations.

Very slow thermal effects can be tolerated provided they are below the band of frequencies to be detected. A large thermal time constant is desirable. Thermal mass will be inherently small in the fiber sensor design. Thermal response time can be increased by increasing the thermal resistance to surroundings. Low-density plastic foams coated with thin reflective films can be used for this purpose.

4.16 Fiber Sensor System Design Considerations

To achieve optimum fiber sensor performance, commercially available components must be analyzed and carefully selected. Fibers, lasers, couplers, connectors, and detectors should be evaluated for their individual suitability to a particular sensor system design. Although the ultimate length of the sensing and reference arms will dramatically affect the system's optical power budget, a 30 dB power dissipation between the laser output and the detector window is a reasonable guideline by which to evaluate the permissible losses through individual components and connectors. As a typical "worst-case" example of a system power budget, consider the following:

Laser diode pigtail output (min): 1.5 mW
Fiber-coupler connector loss (max): –1.0 dB

Splitter excess loss: −0.1 dB
Coupler-fiber connector loss (max): −1.0 dB
Fiber-combiner connector loss (max): −1.0 dB
Combiner excess loss (min): −0.1 dB
Combiner to detector: −1.5 dB

The total is 4.7 dB minus total fiber loss at 4–5 dB km^{-1} = 9.7 dB max, or 160 µW power at the detector, which even with the use of an Si p–n photodiode, leaves a comfortable margin of power.

Several fibers are available commercially with 4–5/80 diameters, single-mode operation at 850 nm, and optimum characteristics for fiber sensor implementation. Although the various epoxy acrylates used as fiber jacket coating are easily removed or chemically thinned accurately, a minimum thickness can be specified which does not require removal for achieving the desired mechanical coupling to transducer materials. Additionally, these coatings provide minimal microbending losses and high survivability in high-strain, high-environmental stress environments. The outstanding feature available in these fibers, however, is the polarization-preserving capability, such that the propagating light is maintained or "locked" into one particular plan of polarization. Polarization preservation is extremely stable under both radial and bending stresses and temperature variations, so that superior fiber-sensing performance and sensitivity are afforded.

Diode laser problems that can result in a loss of sensor sensitivity include amplitude noise, coherence length, phase noise due to small-frequency instabilities, and optical feedback from the load. To avoid excess phase noise, path lengths should be balanced to approximately 1 mm. By using amplitude noise subtraction and ensuring optical feedback levels below 0.001%, 1 µrad performance down to 10 Hz is achievable [1].

Several GaAlAs laser diodes are available that emit at approximately 830 nm a stable, fundamental transverse mode. These devices are characterized by high-power outputs and by linear light output versus current input. Recent advances in diode-to-fiber coupling permit ample power coupling through a single-mode fiber pigtailed directly to the laser. Typical threshold currents are 15–70 mA, with 1–5 mW/facet output at Ith +25 mA, maximum power of 10–15 mW per facet, and beam divergence half angles of $10° × 30°$. If the laser linewidth is very narrow and the frequency does not change, the sensing and reference arms may differ in length by an integer number of wavelengths without affecting the operation except for aggravating the thermal drift problem. However, if the arms differ in length and the light source frequency varies, the frequency shift will be transformed into a differential phase variation and will contribute to the noise from the detector. The phase variation $\Delta\phi$ is $\Delta l/\Delta f$, where Δl is the difference in the optical path length of the arms, and Δf is the variation of the light source frequency.

4.17 Laser Diode Frequency Stability Considerations

Experimental measurements of the output spectrum [18] produced by single-mode semiconductor lasers resulted in observations of instabilities including mode jumping, phase noise, mode partition noise, line broadening, and frequency shifting. Unstable operation in fiber-optic interferometric sensing systems has also been observed. Although stability problems exist for certain applications, semiconductor lasers produce excellent performance in communications and coherent transmission under carefully controlled operating conditions.

In communications, the signal bandwidth can and usually does exceed the laser linewidth [19]. For example, a modern communications system using a directly modulated laser can support a signal bandwidth of 2–3 GHz. This signal bandwidth is much wider than the laser linewidth (a laser linewidth of 10 kHz was reported through the use of a fiber-optic ring resonator). Because of this wide bandwidth, some of the problems associated with frequency instability are not critical to operation in noncoherent communications systems. In sensors and interferometers, however, the signal bandwidth may be very small in comparison to the laser linewidth. For example, if the sensor is designed to monitor ambient room temperature or perhaps the magnetic profile of a slow-moving object, the signal sensed may be in the near dc range, while the linewidth of the laser used in the sensor itself may be in the 1 GHz range. If the laser linewidth suddenly broadens, or changes frequency during operation, a resulting amplitude or phase fluctuation may be produced by the sensor, thus degrading performance due to the "random noise." The need for frequency stability and low noise in the light source for sensor applications is extremely important and may even impose more stringent requirements on laser operation than can be achieved with most devices.

Some laser structures, such as short cavity, external cavity, DFB, and the rare-earth-doped heterostructure, offer various degrees of stability; but in general, the ideal stabilized laser diode source is still in development. Various stabilization techniques such as thermoelectric control of the laser structure, feedback control of the laser injection current, the use of the external fiber-optic ring resonator, etc., are an "after the fact" approach to solving the problem of laser instability, since these methods do not address the source of the problem.

This section is concerned with frequency observations and stabilization techniques applicable to single-mode lasers. Included in the discussions are basic laser operation, the effects of high-frequency modulation and modulation depth on laser operation and received noise, the effects of light reflections back into the laser cavity, stability requirements for an interferometer applications, and laser and linewidth stabilization through the use of fiber-optic resonators and the rare-earth-doped HRO laser.

4.17.1 Laser Operation

Some of the basic requirements for laser operation include the use of direct bandgap semiconductor properties, low photon absorption in the bulk material, high probability of stimulated emission through pumping, the existence of an intense flux of photons at the diode junction, and the existence of an inverted electron population. With these requirements, the semiconductor junction diode will operate as a laser source with the photon amplification being collimated and the emitted energy in the 1–25 Å wavelength range.

For direct bandgap materials such as GaAs, by definition, the conduction band minimum and valence band maximum occur at the same point in "k" space (see Figure 4.17). In this case, only a photon is required to cause an electron transition. For indirect bandgap materials, an electron transition absorbs a photon and absorbs or creates a phonon. The direct transition is more probable, so that direct bandgap materials are more optically active than indirect materials.

Light emission in semiconductors is of two types: spontaneous emission and stimulated emission. With spontaneous emission, a hole and an electron randomly combine. With stimulated emission, a hole and an electron are stimulated to recombine by an existing photon. When an electron absorbs a photon, the electron will make a transition to a higher energy state. When an electron falls to a lower energy state, a photon is emitted. The emission of one photon can cause other electron transitions. The light-emitting transitions occur in the bandgap, and the wavelength of the emitted light roughly corresponds to the absorption edge wavelength of the material. Note that the bandgap decreases with increasing temperature so that the energy of the emitted photon also decreases [20] (see Figure 4.18).

A requirement for an emissive transition is that there must be a filled upper (initial) energy state and a corresponding empty lower (final) energy state, so, to obtain significant light emission, increased concentrations of holes and

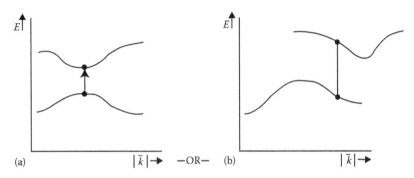

FIGURE 4.17
Direct absorptive electronic transitions: (a) in a direct bandgap material and (b) in an indirect bandgap material.

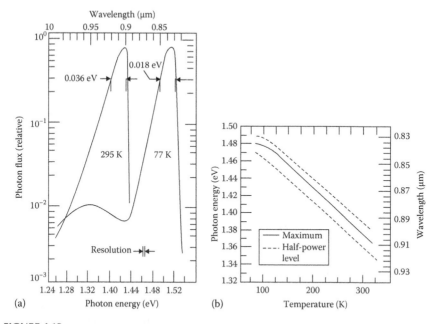

FIGURE 4.18
(a) GaAs diode emission spectra at 300 and 77 K and (b) dependence of emission peak and half-power width as a function of temperature. (After Carr, W.N., *IEEE Trans. Electron Devices* ED-12, 531, 1965.)

electrons must be achieved in comparison to the hole electron concentrations at thermal equilibrium. This increased concentration of carriers, also referred to as an inverted population, can be created in several ways. In the case of a semiconductor laser diode, the process is referred to as pumping.

In a forward biased p–n junction diode, when holes and electrons recombine in the junction region, photons are generated. The light emitted under stimulation by the applied forward bias is called electroluminescence. If light is produced by the injection of photons into the junction region from an external source, the light emitted by the device is a result of photoluminescence. In the case of a light-emitting diode, the light emitted may be heavily reabsorbed in the bulk material outside the junction. Conversely, the laser diode is constructed such that absorption in the bulk material is reduced by making the layer of material between the junction and the external surface thin, and by choosing a material with as low an absorption coefficient as possible. The ability of the light source to produce an emission is rated by the internal and external efficiencies:

$$QE_{ext} = \frac{\text{number of destred photons emitted}}{\text{number of electron} - \text{hole pairs injected}} \qquad (4.4)$$

$$QE_{int} = \frac{\text{number of photons generated}}{\text{number of electron} - \text{hole pairs injected}}$$

The quantum efficiency can also be described as the fraction of the excited carriers that combine radiatively to the total recombination [21]:

$$QE = \frac{R_R}{R} = \frac{T_{NR}}{(T_{NR} + T_R)} \tag{4.5}$$

where
T_{NR} is the nonradiative lifetime
T_R is the radiative lifetime
R_R is the radiative recombination rate
R is the total recombination rate

It is important to note that the quantum efficiency decreases with increasing temperature.

Lasing action occurs when a photon encounters a hole–electron pair at the proper energy separation. When this happens the hole–electron pair recombines and a photon is emitted. The photon emitted will have the same energy, direction, phase, and polarization. The emission will be highly collimated and the spectral width will be in the 1–25 Å range.

4.17.2 Effect of Modulation and Modulation Depth on Mode Spectrum

To begin the discussion assume we have a laser structure that oscillates predominately in a single longitudinal mode when biased at some minimal threshold current level. The threshold current is defined as the minimum current density to obtain the threshold condition. Threshold occurs when the number of photons generated by the structure exactly equals the number of photons lost. Loss mechanisms include scattering loss (loss due to imperfections, voids, or contamination within the lattice), absorption loss (loss that occurs when a photon gives up its energy to an electron or hole in the conduction band or valence band respectively), and emission from the laser.

For a simple p–n junction laser, the current density required to obtain threshold is [22]:

$$Jth = \frac{8\pi e n^2}{(QE)\lambda_0^2}\left(\alpha + \frac{1}{L}\ln\frac{1}{R}\right) \tag{4.6}$$

where
Jth is the threshold current density in Coulombs s^{-1} cm^{-2}
N is the index of refraction

$\Delta\gamma$ is the linewidth of spontaneous emission in s^{-1}
D is the thickness of emitting layer in cm
Q.E. is the quantum efficiency
α is the absorption coefficient
L is the length of active laser structure in cm
R is the reflectivity of Fabry–Perot surface

Note that for a homostructure laser the threshold current increases rapidly with temperature [23].

In the steady state, when the laser is excited with a current and the laser is at or above threshold, the energy of emission will be distributed in various longitudinal modes. The laser may then be excited or modulated by an additional current. The application of the current pulse will cause a redistribution of power in the various longitudinal modes. If there are N longitudinal modes, the amplitudes are defined by [24]:

$$\frac{e^{[-(\lambda_1-\lambda_2)^2/2\sigma^2]}}{\sum_{i=1}^{n} e^{[-(\lambda_1-\lambda_2)^2/2\sigma^2]}} \tag{4.7}$$

where
λ_c is the center wavelength
λ_i is the wavelength of the ith mode
σ is the half width of the spectrum

Even though the energy may be redistributed, the total optical power is constant and may be expressed as follows:

$$\sum_{i=1}^{N} a_1 = 1 \tag{4.8}$$

Depending on the structure of the laser, the redistribution can occur in 0.5–5 ns [25]. Since semiconductor lasers have a very short photon lifetime, modulation at high frequency can be achieved. At the peak amplitude of the combined steady state plus the modulating current, the various longitudinal modes will correspond to the continuous wave spectrum for the combined current, assuming that the laser can respond to the modulating disturbance and that the peak combined current does not exceed device limitations.

During switch on of the laser, the ratio of power in the ith mode to that in the jth longitudinal mode can be expressed (for pulsed or step excitation, neglecting spontaneous emission) as follows [26]:

$$\frac{S_i(t)}{S_j(t)} = \frac{S_t(t=0)}{S_j(t=0)} e(G_i - G_j)_t \tag{4.9}$$

where

  $G_i = g_i\alpha$ is the optical gain of the ith mode
  g_i is the Lorentzian distribution $= 1(1 + b_i^2)$
  α is the optical gain of mode zero
  b is the mode selectivity of the ith mode

For a laser structure with a gain spectrum of several hundred Angstroms, the mode selectivity is in the range of 10^{-4} so that it takes a relatively long time for the different modes to settle to their steady amplitude. Assuming α occurs at the peak of the optical gain curve and that spontaneous emission is negligible, then at $t = \infty$ only one mode can exist because all other modes will decay to zero. The above equation cannot be used to describe the time-dependent spectrum if the laser is modulated continuously at high frequency. It has been observed that the lasing spectrum under high-frequency (microwave) modulation will remain single mode if a critical modulation depth is not exceeded [27].

4.17.3 Experimental Observations

Following are some important observations concerning mode stability in semiconductor laser diodes.

- Carrier density fluctuations that occur under certain conditions of high-frequency modulation result in increased linewidths of the individual lasing modes [28].
- The spectrum of direct modulated lasers depends on the amount of mode selectivity in the laser.
- Lasers can be constructed with a frequency selective element. This frequency selective element is used in the DFB laser. The DFB laser can sustain single-mode oscillation even under turn on transients and high-frequency modulation [29].
- Lasers having a very short cavity length sustain single-mode oscillation under turn on as well as under high-frequency modulation [30].
- Short cavity and composite cavity lasers settle to a single mode faster than a long cavity laser of similar construction [31].

4.17.4 Guided Index and DFB Laser Operation

Many critical applications such as optical disc reading and recording or coherent communications require a stable single-mode laser. One method presently used to achieve single-mode operation involves the use of a controlled perturbation in the plane of the p–n junction [32] to act as an internal waveguide to control the mode of operation. This structure is known as the index-guided laser. To provide a controlled frequency of operation, that is, linewidths in the 0.5–0.8 Å range, the laser structure can be fabricated to incorporate a grating for DFB. A typical grating spacing is in the 3450–3476 Å range. The grating is constructed in such a way that it suppresses unwanted wavelengths while selecting a chosen wavelength. The wavelength selective grating is used in place of facets, which are not frequency selective, to provide the optical feedback necessary for lasing. In the DFB laser, the wavelength of emission is somewhat affected by the laser gain curve [33] and more specifically by the grating spacing. Single-mode operation depends on the fact that the frequency spacing between adjacent modes is large and the grating can significantly reduce the gain at the unwanted frequency.

DFB laser structures that operate in the 1.3–1.55 μm range are commercially available. Very short grating spacing is required for operation around 0.83 μm, and developments are more limited due to manufacturing difficulties. Lattice damage often occurs during the masking or chemical etching or ion beam sputtering grating fabrication processes, lowering the optical quality of subsequently grown layers.

Figure 4.19 shows the CW characteristics of a 120 μm buried heterostructure GaAs laser [34]. Some observations include the low threshold current and the single-mode spectral output at an optical output of 1.3 mW. The characteristics of a similar 250 μm buried heterostructure GaAs laser are shown in Figure 4.20.

Note that a higher threshold current is required and that single-mode operation was not achieved. Figure 4.21 depicts the time-averaged spectrum

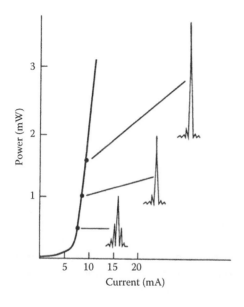

FIGURE 4.19
CW light versus current and spectral characteristics of a GaAs laser whose cavity length is 120 μm.

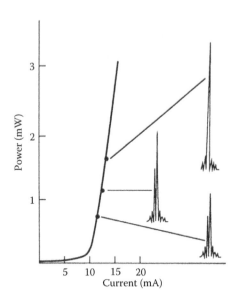

FIGURE 4.20
CW light versus current and spectral characteristics of GaAs laser identical to that
shown in Figure 4.19 except that the cavity
length is 250 μm.

at various modulation depths at modulating frequencies between 1 and
3 GHz. Note that a single-mode spectrum is achieved in the laser with a
125 μm cavity length at modulation depths up to 90% independent of modulating frequency, but multimode oscillation occurs at 75% modulation
depth independent of modulation frequency for the longer cavity laser. The
linewidths of individual modes will broaden as the frequency of modulation increases but the amplitudes do not change. This can be explained as
a result of fluctuations in the cavity refractive index because of fluctuating
carrier density, that is, under constant optical modulation depth, the fluctuations in carrier density increase with increasing modulation frequency,
so that line broadening is more visible at higher modulating frequencies.
For a homogeneously broadened gain spectrum (no mode jumping), when
the laser is biased to a certain optical power and is modulated at a high
frequency and the modulation depth controlled, the time averaged lasing
spectrum is equivalent to that of a laser-operating CW at a reduced power
level [35].

4.17.5 Modulation Depth and Signal-to-Noise Considerations

The emission spectrum of a single-mode laser under continuous microwave
modulation will exhibit a single longitudinal mode if the modulation depth
generally does not exceed 80%. If the modulation depth is increased above
80%, the lasing spectrum will begin to break into multimode oscillation. The
spectral dynamics can be estimated by observing the CW lasing spectrum at
various output levels. Another requirement for stable operation is to maintain the laser above threshold throughout modulation.

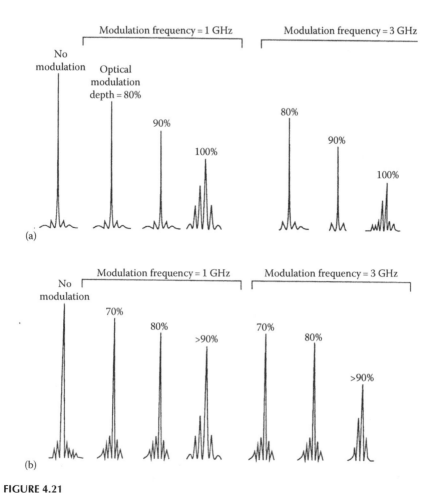

FIGURE 4.21
(a) Observed time-averaged spectrum of the laser shown in Figure 4.20 under microwave modulation at various optical modulation depths, at modulation frequencies of 1 GHz. The laser is biased at a dc optical power of 1.5 mW. (b) Same experiment as in (a) but for the laser shown in Figure 4.21.

It would appear that stable laser operation can be ensured by increasing the bias, but other undesired effects occur with increasing bias. These effects include temperature instability and line broadening due to the effect of light reflected into the cavity from external sources. The temperature instability results mainly from shifts in the bandgap with temperature, whereas the line broadening results when intensity within the laser cavity is increased above a critical level.

The method described for maintaining single-mode operation through reduction in the modulation depth may in some cases result in satisfactory system performance; however, evaluation of the overall system performance must include consideration of the many noise sources. Some of these noise sources include phase noise and noise introduced through the modulating signal itself. This provides a foundation for reducing emission intensity or modulation depth to maintain laser stability.

In the optical receiver, the noise current is the sum of quantum noise, thermal noise, dark current noise, leakage current noise, and beat noise. Beat noise is zero in a single-mode system, while the thermal noise is usually dominant. Modulation levels must be selected to ensure a satisfactory distortion level, with the tolerable level depending upon the nature of the information being transmitted.

4.17.6 Instability due to Optical Feedback from Distant Reflectors

Instability can occur as a result of the finite phase and carrier number change caused by fluctuations in spontaneous emission [36]. This instability only occurs when the laser reaches a steady state that maximizes coherent feedback and laser light intensity. The instability caused by reflections vanishes at strong feedback levels if the reflected energy is in phase with the emission photon. The laser is nearly stable at threshold with moderate feedback but unstable when operated well above threshold.

4.17.7 Stability with Moderate External Feedback

For the following discussion we assume a laser system operating in the presence of a moderate level of external feedback (−30 dB) from some component many cavity lengths away from the laser active region. These represent the typical conditions for applications such as video recording, coherent communications, and sensing where external modulation is utilized [37]. The effects of reflections on a CW laser are well documented [38]. Weak reflections cause line narrowing [39] and high levels of reflection cause line broadening [40]. With moderate levels of optical feedback, the laser line becomes significantly broadened to a linewidth on the order of 25 GHz. This compares to a linewidth of from one to several Angstroms for a stable laser operating in steady state with zero feedback. The effect of the excessive line broadening (coherent collapse) is attributed to a lack of coherence between the field of the continuously operating laser and the steady-state reflected field.

Under moderate feedback, a CW laser will exhibit periodic low-frequency intensity fluctuation [41]. The dependence of the period of the fluctuation, reflector separation, and laser current are related. The pulsation period is related to the time required for a photon to make approximately ten round trips in the external cavity and thus is related to cavity length. When the laser is operating near threshold, the intensity of the pulsating light is amplified because of the many round trips in the external cavity. Between pulsations, the intensity increases and approaches a steady-state intensity, then suddenly drops to zero and the process repeats.

Line broadening of up to 40 GHz occurs when the laser is brought from a condition of below threshold to threshold lasing operation. At one point in the threshold, the gain is a minimum and the pulsations are strongly damped, decaying to zero. The pulsations occur when the laser is operating above threshold and the cavity exhibits gain [42]. Without external feedback, as the carrier number increases, the gain increases. This increase in gain results in an increase in light intensity, thereby increasing stimulated emission. The increase in stimulated emission in turn decreases the carrier number. The effect leads to a relaxation oscillation. The relaxation oscillation is usually damped by other effects.

Instability arises because the laser seeks a steady-state operation that maximizes feedback. For example, the maximum feedback state occurs when the stimulated emission is in phase with the reflected field. An increase in the number of carriers occurs within the cavity, modulating the intensity and the phase difference, and decreasing the intensity. Eventually the fluctuations in carriers and phase become greater than the restoring forces and instability occurs. In steady state with moderate feedback, the laser is phase locked to the reflected field.

4.17.8 Laser Frequency Stability Considerations in Fiber-Optic Sensors

Fiber-optic sensing is a technology area in which applications continue to emerge at a rapid pace. The sensing mechanisms in a fiber sensor can include one or all of the following modulation techniques: intensity, phase, polarization, or frequency. The sensitivity of fiber sensors makes them attractive in many applications where electronic sensors are inadequate. The interferometric class of fiber sensors offers the highest sensitivity and dynamic range. The parameters to be sensed cause a change in the phase of the interferometer, eliminating the need for a separate modulator [43].

The major difficulty in implementing the interferometric fiber sensor involves the requirement for a coherent and stable light source. Accuracy of sensing is directly impacted by the laser source and any mode jumping, phase noise, or mode partition noise that occurs. Laser structures with the most stable operation possible are chosen for the fiber sensor system. For applications in which size and power consumption are not limited, a gas laser should be considered since the gain curve of the total linewidth is typically on the order of 1 GHz [44]. With the gas laser, it is relatively easy to define the wavelength to parts in a million; in some cases, the HeNe laser can be locked to the center-gain curve with the stability on the order of parts in 100 million [45]. This parameter is very important because changes in the operating frequency result in time delays due to fiber dispersion, even in the low dispersion operating regions of the single-mode fiber.

Along with the effects of dispersion and frequency instability, the performance of the sensing system can be affected by the presence of phase noise. Phase noise occurs due to imperfections in the laser cavity, laser drive circuit noise, and resonance within the laser structure itself [46]. Some control of excess phase noise is possible through implementation of a feedback control system in the laser drive. Tight control of the laser drive current is necessary because the laser resonance is a function of the drive current. Typically, a change in drive current of 1 mA will result in a several gigahertz change in lasing frequency. Other enhancements such as modification of the laser emission linewidth through the use of an external cavity and the use of thermoelectric cooling of the laser body provide only minimal improvements. Laser source stability improvements are necessary for some of the emerging applications in fiber sensing. Further developments are also necessary before laser manufacturers offer the advanced laser structures in the wide range of needed operating frequencies; therefore, external stabilization are employed as an interim solution. One such approach to achieving stable operation of the laser source involves the use of an external electronic feedback circuit and the fiber ring resonator.

4.17.9 Achieving Laser Stability through External Control

The external cavity structure was developed for reduced linewidth and improved stability, but these structures are still susceptible to instability due to thermal and mechanical influences on the optical feedback and laser operation [47]. These temperature stability and mechanical effects are a direct result of the external structures being spread over a wider surface area, which is more vulnerable to external influence. The system shown in Figure 4.22 was

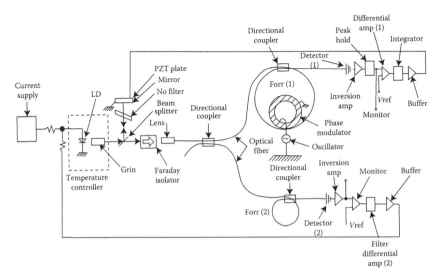

FIGURE 4.22
Schematic configuration of experimental arrangement for the stabilization of spectral linewidth and oscillation frequency of the external-cavity laser diode.

demonstrated by Tai et al. [48] to offer a stable spectral linewidth of 2 GHz and a stable frequency of operation with fluctuations less than 2 MHz p–p. The linewidth was stabilized through the use of automatic phase control, while the laser oscillation frequency was controlled through the use of automatic injection current control. The basic system shown in Figure 4.22 includes the following components:

- Laser diode: A GaAs/AlGaAs laser operating at 830 nm and threshold current of 60 mA. The laser is temperature stabilized with a thermoelectric element to 13°C ± 0.01°C.
- Mirror: A mirror mounted on a PZT plate 15 cm from the laser is sued to provide controlled optical feedback. Feedback is also further controlled through the use of a 1 m fiber-optic ring resonator.
- FORR(1) and FORR(2): A fiber-optic ring resonator constructed with low loss (0.2 dB) single-mode fiber. FORR(1) also includes a PZT phase modulator.
- Directional couplers: Directional couplers fabricated by the polishing technique [49].

Linewidth control is provided by the control of the mirror position and FORR(1). The resonance peak detected by the photodiode(1) and the reference are compared by the differential amplifier. The output of the differential amplifier is further processed by an integrator and a buffer amplifier. The output of the buffer amplifier is then used to control the mirror position by exciting the PZT plate upon which the mirror is mounted. The cutoff peak of the FORR(1) and control loop is 1 kHz.

Frequency stability is provided by FORR(2) and the lower control loop. Frequency fluctuation in the ±12 MHz range is converted into FORR(2) transmission intensity. The output signal is detected by photodiode(2). The voltage produced by photodiode(2) is further processed by comparing this signal with a reference voltage at the differential amplifier and filtering via a 10 kHz low-pass filter. The buffered output of the filter is used to control the laser diode injection current.

The major drawbacks of this system are the need for external components and electronic circuitry. The use of many external components must be weighed against the use of a gas laser where external components may not be necessary to achieve stable operation.

4.17.10 Rare-Earth-Doped Semiconductor Injection Laser Structures

In an ongoing effort to improve the stability of laser diode operation, manufacturers are working to design structures that provide emission frequency stability over a wider temperature range, provide a decreased linewidth, eliminate mode hopping, provide a reproducible center frequency from device to device, and provide a structure in which emission is immune to external reflections. The rare-earth-doped laser device may meet all of these requirements.

This type of current injection laser was reported [50] to provide a stable single longitudinal mode of operation in the near-infrared 1.55 μm wavelength. The device uses a rare-earth erbium dopant in the active layer, chosen so that the emission wavelength of 1.5322 μm from the trivalent rare-earth ion is shorter than the band edge emission of the host semiconductor at 1.55 μm. This results in a narrow spectral emission superimposed on the broad gain peak of the host semiconductor (see Figure 4.23).

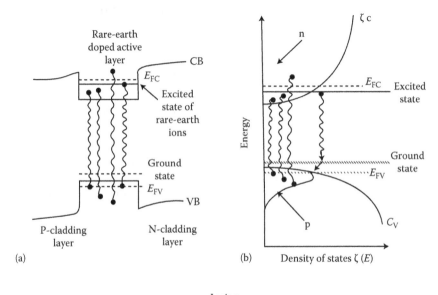

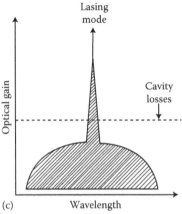

FIGURE 4.23
Rare-earth-doped semiconductor injection laser structure. (a) bandgap diagram (b) energy densities (c) gain curve.

The device was fabricated with a heteroepitaxial ridge overgrown (HRO) construction. The structure uses conventional Fabry–Perot cleaved facets, a cavity length of 250 µm, standard cleaving, and standard ohmic contacts. The 5% by weight Er doping was added to the In melt during the liquid phase epitaxial growth of the layer. The Er diffuses into neighboring layers during the growth process. The HRO laser operated in a single longitudinal mode, was stable with temperature, had line shifts of less than 1 Å C^{-1}, was apparently immune to external reflections, and showed reproducible lasing frequency in devices from different wafers.

The concept of using rare-earth doping has been used in the past. For example, rare-earth ions were incorporated into III–V structures [51] and silicon [52] by ion implantation and other doping methods. At low temperatures, that is, 77 K, the emissions from the rare-earth samples did not depend on the band-gap energy of the host semiconductor but obeyed the emission expected of the particular rare-earth ion. It is also common practice to use rare-earth ions to produce emissions having a wavelength longer than the band edge wavelength of the host semiconductor. For example, in light-emitting diodes, rare-earth ions are used in the "up-converter" [53] where the infrared emission from a GaAs light-emitting diode is absorbed by phosphor doped with rare-earth ions such as ytterbium (Yb^{3+}) and erbium (Er^{3+}). The operation is dependent upon the successive absorption of a single photon in the visible region.

Figure 4.24 shows a comparison of the spectra for an Er-doped HRO laser and a control non-rare-earth-doped HRO laser at different injection levels. Note that the control HRO laser shows typical multilongitudinal mode operation including mode hopping that is common with quaternary lasers, while the Er-doped laser exhibits a clean single longitudinal mode of operation without mode hopping. These initial results are very encouraging, and continued development will lead to increased commercial availability of these devices.

4.17.11 Solutions to Laser Frequency Instability: Summary

Some of the viable solutions to the problem of laser frequency detailed in the previous sections are as follows:

- Maintain the laser above threshold during operation. If the laser is biased substantially above threshold, single-mode operation can be maintained at a higher level of modulation than when the laser is biased close to or below the threshold level.
- Operate the laser at an 80% modulation depth maximum. High-frequency modulation has little effect on the lasing spectrum if the modulation depth is maintained at or below 80%.
- Operate the laser at an intensity level that ensures stable operation.
- Reduce cavity length or use a laser source that includes an external cavity.

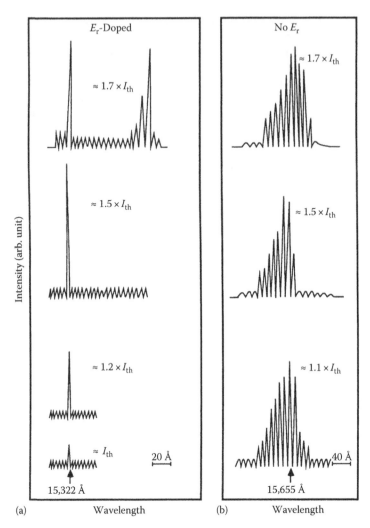

FIGURE 4.24
Comparison of the lasing spectra for (a) an EA-doped HRO laser and (b) control (no Er) HRO laser at different current injection levels.

- Use an antireflection facet if reflections cannot be avoided.
- Rare-earth doping of the heteropitaxial ridge overgrown layer shows encouraging results.
- Automatic phase control and automatic current control can be used to achieve reduced linewidth and stable frequency of operation.
- Maintain a constant laser temperature during operation using thermoelectric control.

4.18 Fiber Sensor Design Example: Fiber-Optic Sonar Dome Pressure Transducer

4.18.1 Identification and Significance of the Problem

Influence-actuated sensors and mines require means of detecting and classifying influences produced by vessels. Among these are pressure fluctuations produced by the bow wave, propeller cavitation, and engine noise, and magnetic field disturbances produced by the presence of the steel vessel in the earth's field. These influences exist in a noise field created by wind and tide driven surface wave, sea creatures, distant sources propagating in density gradient guides, and other out of range shipping.

Reliable sensing often depends upon correlation of all influences with individual signals weighted according to their reliability and expected SNR. More sensitive detectors are needed. Fiber-optic interferometric sensors possess very high sensitivities and dynamic ranges. Techniques exist for applying them to each of the signal types mentioned above. Acoustic and seismic sensors can also be made directional so that the bearing angle can be determined from an array of three sensors. If the sensors are adequately separated, range can be deduced from the time difference of arrival of the sound. Spectral characteristics of the various influences can be correlated with known target properties and used for target classification.

4.18.2 Possible Solution for a Sonar Dome Pressure Transducer

The Mach Zehnder fiber interferometer is the most convenient and most frequently used fiber sensor configuration. Figure 4.25 illustrates this form schematically. It consists of a coherent laser source whose output is split into two equal paths by a 3 dB directional coupler. The two light signals are transmitted through separate equal length fiber paths and recombined in a second 3 dB directional coupler. The second coupler outputs are converted to electric signals by photodetector diodes. If the two fibers are identical in length and are made from identical materials, the two light signals arriving at the second coupler will be in the same time phase and will add constructively to produce a maximum output intensity.

If one fiber is one half wavelength longer than the other, the two light signals will be of opposite time phase and will add destructively to produce a minimum output intensity. This is true irrespective of the total lengths of the fibers; hence, an elastic elongation of one fiber by one half wavelength will produce a maximum output variation. The system can be viewed as a very sensitive strain gage measuring displacements in terms of the wavelength of light, with typical phase resolutions on the order of 1.0 μrad in a 1 Hz bandwidth.

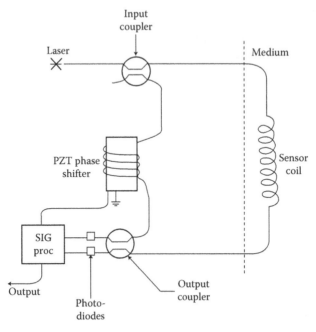

FIGURE 4.25
Interferometric sensor concept.

As a practical matter, if the unperturbed sensing fiber and the reference fiber are exactly the same length, the combined outputs will be a maximum, and either shortening or lengthening the sensor fiber will produce a reduced output. It is therefore not possible to discern from the output whether the quantity being measured is initially increasing or decreasing. Also, the rate of change in the magnitude of the sum of two sinusoids with respect to phase offset is very small when the phase separation is small.

Both of these difficulties can be overcome by inserting a fixed quarter wavelength (90°) phase difference between the sensing and reference fiber paths. In this case, the peak of one waveform occurs at the zero of the other, so that the sum is one half of the maximum occurring when the phase difference is zero. The slope at the zero crossing is maximum and the sums change rapidly. The transfer matrix of the directional coupler is such that when the path lengths differ by 90°, the two outputs to the photodetectors are of equal magnitude.

The detector outputs are applied to a difference amplifier whose output is then zero in the undisturbed state. When the phase difference is zero, the sum of the amplitudes is maximum at one photodetector and zero at the other. When the difference is 180°, the condition is reversed; hence, the difference output is zero for the undisturbed state and goes positive or negative depending on which fiber is longer. Since the wavelengths in use are on the order of 1 μm (0.000039 in.), the system is in effect an ultrasensitive strain gage. Any measure and that can be converted to a fiber strain can be measured with comparable sensitivity. If one arm is spatially displaced from the other, the system becomes a gradiometer sensitive to the spatial derivative of the measure and field.

Pressure variations can be converted to axial fiber strains by wrapping the fiber around a pressure expandable mandrel or by jacketing the fiber with an elastic plastic or rubber material. Figure 4.26 shows the latter. Here an increase in ambient pressure causes the jacket to elongate in one direction. This strain is transferred to the fiber, increasing its optical phase length. The ultimate resolution of the fiber-optic interferometer is determined by the smallest detectable phase difference between the light waves emerging from the separate paths. The literature reports a noise-limited maximum phase resolution of 1 μrad under a standardized set of reference conditions that control other variables affecting sensitivity. These conditions are as follows:

- A signal bandwidth of 1 Hz.
- A difference in the mechanical length of the two fibers of 2 mm or less.
- Two photodetectors are used with a difference amplifier as shown in Figure 4.25.
- The signal frequency is 2 kHz or is an envelope on a 2 kHz carrier.

A few words about these conditions are in order. The 1 Hz bandwidth is a convenient reference, and the SNR or limiting sensitivity of any other signal bandwidth can be scaled directly. A difference in the mechanical lengths of the fibers will make the system output a function of the laser wavelength.

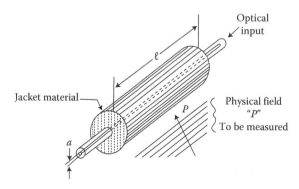

FIGURE 4.26
Acoustic pressure transduction mechanism.

In order to avoid detection of laser FM noise and frequency drift, the length difference must be minimized. A practical limit is imposed by fiber handling and dressing procedures. Experience shows that attempting to match lengths closer than 1 or 2 mm is unproductive and unnecessary. Use of balanced differential detectors minimizes common mode effects caused by laser amplitude fluctuations.

Referencing the resolution measurements to a 2 kHz signal frequency is one approach to avoid some of the effects of $1/f$ noise present in the laser and electronics. The anticipated signal frequencies are much less than 2 kHz. The simple homodyne detection scheme shown in Figure 4.25 is subject to $1/f$ noise; hence, the sensitivity will be diminished by noise. We will show that more than adequate sensitivity can be achieved. Some additional areas where component improvements will benefit system performance are as follows:

- Source diode
  - Stabilized diodes
  - Reduction of $1/f$
  - Long life
- Integrated optics
  - Low cross talk
  - Low loss
  - High efficiency
- Polarizers
  - High extinction ratio
  - Availability of single polarization fiber
- Couplers
  - Extra low loss
  - High birefringence
- Modulators
  - Frequency shifter
  - High frequency, high efficiency
- Fibers
  - Low loss, high birefringence
  - Low loss polarizing

4.18.3 Feasibility Analysis

To determine feasibility, we begin with the noise-limiting value of 1 μrad and degrade it by the difference in $1/f$ noise between 2 kHz and 1 Hz signal frequencies. By this we are saying that the smallest detectable phase difference is 2000 µrad. We can define our goal to measure with good precision a pressure change of 0.01 in. of water. For reasonable accuracy this should be equivalent to 10 times the smallest resolvable phase increment. This means that a phase change of 20,000 µrad will represent a pressure change of 0.01 in. of water. Since phase change relates directly to fiber elongation and optical wavelength, we can pick an appropriate light source and calculate the necessary elongation as a function of pressure without initially defining the pressure transduction process.

Selection of the light wavelength is influenced by the availability and cost of suitable single-mode laser diodes and single-mode fibers. Reduction of wavelength translates to smaller fiber diameters and lower losses for a given bending radius. However, intrinsic fiber losses are higher for the shorter wavelengths. Low loss transmission windows at 1.3 and 1.55 µm provide a strong incentive for operation at those wavelengths. Suitable laser diodes are available at 830–850 nm as well as at 1.3 and 1.55 µm. We will use the 1300 nm wavelength for feasibility calculations. The actual fiber elongation is nearly

$$\Delta l = \frac{20,000 \times 10^{-e}}{2\pi} \text{(wavelengths)} \times (1.3 \times 10^{-6})\text{m}$$

$$= 4.14 \times 10^{-9}\text{m} \tag{4.10}$$

This is the elongation to be produced by a pressure increase of 0.01 in. of water. Particular applications may require a dynamic range of say ±20 in. of water. For this case, the total elongation will be

$$\Delta l = \frac{2(20)4.14 \times 10^{-3}}{0.01}$$

$$= 16.56 \times 10^{-6} \text{ m} \tag{4.11}$$

We need to find the total length of fiber in each branch. The minimum overall length is related to the total elongation and the allowable stress in the fiber as well as to the transducer area exposed to the hydrostatic pressure. Polarization-preserving single-mode doped silica fibers are the appropriate choice. The tensile stress at which a fiber will fail depends on the composition, residual internal stresses, and surface defects. Polarization-preserving birefringence is induced by the inclusion of stress rods in the fiber when it is fabricated. For these reasons, the ultimate allowable stress is considerably less than could be expected from the properties of pure silica alone. We used ITT fiber #T1605 in previous related experiments [54]. The manufacturer's proof test in this case is 25,000 psi. If we assume a 10:1 safety margin, the maximum allowable stress is 2500 psi. The elastic modulus is 10.5×10^6 psi and the unjacketed diameter is 80 μm.

Knowing the allowable stress and the desired strain, we can calculate the total length of fiber. The elastic modulus M is defined as

$$M = \frac{\text{stress}}{\text{strain}} = \text{stress} \times \frac{1}{\delta l/i} = \text{stress} \times \frac{i}{\delta l} \tag{4.12}$$

where
  l is the total length
  δl is the total elongation

Then

$$l = \frac{M\delta l}{\text{stress}} = \frac{(10.6 \times 10^6)(16.56 \times 10^{-6})(39.37)}{2500} = 2.74 \text{ in. minimum} \tag{4.13}$$

The tensile force is the stress multiplied by the cross-sectional area.

$$F_T = 2500 \left[\frac{(80 \times 10^{-6})39.37}{4} \right]^2 \pi = 0.019 \text{ lb} \tag{4.14}$$

We based these calculations on the assumption we want to minimize the amount of fiber. The length we calculated is unreasonably short. It will be more convenient to use a longer fiber. This can be accomplished by simply reducing the maximum stress. We conclude that there is no basic problem in achieving the necessary sensitivity.

4.18.4 System Sensitivity

To this point, we related a pressure scale to fiber strain, stress, and tensile force, but did not show the transduction from hydrostatic pressure to fiber tensile force/stress. As mentioned previously, two methods have been employed. One is to wind the fiber around a compliant cylinder with the pressure differential applied between the inside and the outside of the cylinder. The second makes use of the pressure deformation of an elastic jacket applied directly to the fiber. We will examine both methods.

The density of sea water at 15°C is 63.99 lb ft^{-3}. One inch of water is equivalent to a pressure of $P = 63.99/12^3 = 37(10^{-3})$ lb in.$^{-2}$. If the hydrostatic pressure operates on area A, the force is $F = PA$. From the above data, $A = 0.01$ $9 \times 12^3/63.99 = 0.513$ in.^{2}. Again, this figure corresponds to an impractical minimum length of fiber. If the fiber length is increased and the maximum

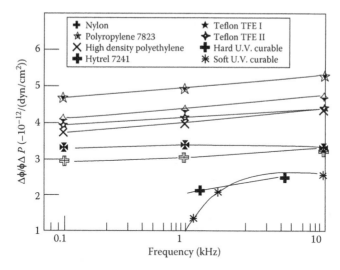

FIGURE 4.27

Frequency dependence of pressure sensitivity of fibers with a 0.7-mm outer diameter coated with various elastomers at 27°C.

stress decreased, the pressure interaction area per fiber strand will decrease accordingly.

Giallorenzi et al. [55] published various phase versus pressure sensitivities for 0.700-mm-diameter fiber jackets of various materials. These data also show the acoustic frequency dependence. Of the several materials shown, Teflon TFE has the greatest sensitivity, but nylon exhibits the smallest frequency dependence. Since they differ only by a factor of 1.4, nylon appears to be the better choice. From the published graph (Figure 4.27), the phase response

$$\frac{\Delta\varnothing}{\varnothing\Delta P} = \frac{3\times10^{-12}}{1.45\times10^{-5}}\frac{\text{rad}/\text{rad}}{\text{lb}/\text{in.}^2} = 214\times10^{-9}\ \text{rad}/\text{rad}/\#/\text{in.}^2 \qquad (4.15)$$

Here $\Delta\varnothing$ is the optical phase change for the pressure change of 1.45×10^{-5} lb in.$^{-2}$ over the length of a fiber whose end-to-end transmission phase delay is $\varnothing$. We now calculate the fractional phase change for a pressure change of 0.01 in. of water.

$$\frac{\Delta\varnothing}{\varnothing\Delta P = 0.01} = 214\times10^{-9}(10^{-2}) = 214\times10^{-11}\ \text{rad}/\text{rad} \qquad (4.16)$$

We want $\Delta\varnothing$ for 0.01 in. of water to be $20{,}000\times10^{-6}$ rad, so we need to multiply $\varnothing$ by the ratio $(20{,}000\times10^{-6})/(214\times10^{-11}) = 9.35\times10^{6}$, that is, $\varnothing = 9.35\times10^{6}$ rad. If we simplify the problem by assuming that the fiber length is the phase length in wavelengths divided by the refractive index, that is, neglecting higher-order effects, we obtain

$$L = \frac{9.35 \times 10^6}{2\pi} \times \frac{1.3 \times 10^{-6}}{1.458} (39.37) = 52.24 \text{ in. minimum} \qquad (4.17)$$

As in the previous case, we have shown that the requisite sensitivity and resolution should not be a problem. This is in keeping with the Giallorenzi statement "The phase (or interferometric) sensor, whether for magnetic, acoustic, rotation, etc., sensing, theoretically offers orders of magnitude increased sensitivities over existing technologies. In the case of the acoustic sensor constructed utilizing optical fiber interferometers—these theoretical predictions have been verified to the limit of state of the art in acoustic measurements."

The Mach–Zehnder interferometer output is proportional to the squared sum of the light amplitudes from the sensing and reference branches and varies as the phase difference between the electric fields varies. The individual amplitudes should remain constant. In order that the output variation be maximum and approach a linear variation with phase angle, a displacement of $\pi/2$ radians or 90° will be created. The operating range of phase variation is on the order of $\pm\pi/4$ rad (45°).

This is much less than the anticipated phase variation of the pressure sensor. Large variations are accommodated by means of a phase-locked feedback arrangement that maintains near zero phase difference at the output by stretching the reference fiber to match the phase delay in the sensing fiber. The control signal to the stretcher in the reference branch is then the system output. For example, a pressure sensor dropped from the surface to a depth of 600 ft would have a large phase-locked loop stress built up by the time it came to rest. Since the absolute depth change during emplacement is of no particular significance, the sensor could be activated after emplacement. In the foregoing calculations, we set the phase shift equivalent of 0.01 in. of water to 20,000 μrad or 20 mrad. If a stable sensor has a more or less linear range of $\pm\pi/4$ radians, this corresponds to pressures of ±78 in. of water. Under ordinary operation, the loop stress would be small, greatly reducing the probability of breaking phase lock.

The bandwidth of a fiber-optic pressure sensor is determined by the resonances of the transducer and the frequencies of environmentally induced drifts. The sea should be a relatively benign environment with regard to thermal effects. It may ultimately be possible to extract pressure, seismic, and acoustic data from a single sensor by spectral filtering.

There are several noise sources that cannot always be distinguished from a pressure signal. These effects and the methods of correction are discussed below. Note that the fiber interferometer discussed in this example could also be used to sense magnetic field disturbances by mechanically coupling a magnetoresistive material to the sensing fiber. Allied Corporation's METGLAS 2605SC is a suitable material whose properties can be tailored to this application [56].

- Unequal thermal expansion: match thermal coefficients of expansion and subject sensing and reference arms to the same thermal environment by thermal filtering
- Transmission loss variation: mount fibers to minimize externally induced stress
- Depolarization, depolarization shifts: use polarization-preserving fiber and minimize radial stresses, temperature, and environmentally induced stresses
- Laser frequency noise: make reference and sensor fiber lengths as nearly equal as possible, stabilize laser frequency with current regulators and temperature controller, and isolate laser from optical feedback

4.18.5 Light Source

Single-mode fiber pigtailed laser diodes are available at low cost that produce power levels of milliwatts at the fiber end. The dynamic (slope) resistance is low at the operating point, and the device is operated at constant current. Light is proportional to current, but the current is temperature dependent.

Typical lasers operate at currents between 55 and 80 mA and at voltages between 1.6 and 1.7 V. Of interest is the fact that lasing does not start until the diode current reaches 55 mA. At this threshold, the input power is approximately 88 mW. At the 1 mW optical power point, the power input is 102 mW. Efficiency would be much better if the threshold current could be substantially reduced. Experimental diodes are reported with threshold currents of 1 mA.

Many diodes have a built-in photodetector that receives a sample of light from the diode facet opposite the output facet. The photodetector output is used to drive an external feedback loop that regulates the laser current to maintain a constant light intensity. Peltier thermoelectric coolers are incorporated in some laser diode packages along with thermistors so that closed loop temperature control can be effected. These coolers require substantial currents (up to 800 mA). They are probably unnecessary in an underwater device where excess heat can be dissipated to the water and where temperature extremes are limited. Underwater temperature changes are expected to be relatively slow in comparison to sensor response times.

4.18.6 Photodetectors

The photodetector diodes must be matched in spectral response to the laser wavelength selected and have a low noise output. The response time must be adequate for the sensing frequencies desired. A convenient arrangement incorporates the photodiode and a low noise operational amplifier in the same case. This reduces noise and lessens the chance of spurious oscillation or instabilities. These detector amplifiers are available in a variety of packages. The active area of the detector is much larger than the fiber cross sectional area, so attaining efficient optical coupling between the fiber and detector is easily achieved.

4.18.7 Single-Mode Fiber Directional Couplers

These are 3 dB hybrid quadrature power splitters that are optical equivalents of the microwave waveguide directional coupler. They are formed by reducing the cladding thickness on two fibers so that the fibers can be placed into close proximity. A common approach for this is to fixture the fibers so they can be twisted and drawn axially while heated. Coupled power is monitored during the process. The resulting fiber structure is quite fragile and must be encased before removal from the fixture. Typical commercially available pigtailed couplers are 1.22 in. long × 0.138 in. diameter.

4.18.8 Optical Fibers

Low loss fibers fall into the following categories: multimode, single mode, single-mode polarization preserving, and single-mode single polarization. Mulitmode fibers are unsuitable for interferometric phase measuring systems because various modes propagate at different velocities, tending to smear any phase information present when the modes are combined at the detector. Ordinary single-mode fibers support two identical TE_{11} (first order) modes orthogonally polarized to each other. If the fiber geometry and refractive index distribution are not absolutely circularly symmetric about the fiber axis, waves of a certain polarization will advance on waves of some other polarization. A linearly polarized wave at some arbitrary angle can be resolved into two orthogonal components. If one of these components advances or is retarded with respect to the other, the polarization will become circular or elliptical depending upon whether the components are equal in amplitude.

Small variations in core-cladding geometry, composition, and stress distribution in an ordinary single-mode fiber will produce this sort of depolarization— the magnitude of which varies with temperature, fiber bending, and unequal radial stresses. Ordinary single-mode fibers are used for long distance telecommunications applications, in which depolarization is not a concern. Single polarization fibers are designed such that they only trap and propagate light of a particular polarization. Orthogonally polarized light leaks from the fiber and is lost.

Polarization-preserving fibers are fabricated to produce a marked difference in propagation velocity between a particular linear polarization and one orthogonal to it. The positions of these fast and slow polarizations in the cylindrical coordinate system of the fiber are called the axes of birefringence. If a linearly polarized wave is launched with its polarization vector parallel to one of these axes, it will retain this polarization over long distances. If it is launched with the polarization vector at 45° to the axes, the polarization will change cyclically from linear to circular to linear, etc., as it propagates along the fiber. The distance along the fiber for the polarization to complete a cycle and return to its original condition is called the fiber beat length. A condition for the effective use of these fibers is the ability to locate the axes of birefringence and to match them when joining the fibers for coupling fibers to sources.

The field amplitudes of the waves emerging from the two fibers of the interferometer add vectorially in the output couplers, and the photodetectors produce an output voltage proportional to the intensity, which is a constant times the amplitude squared. Some of the difficulties involved in transmitting light into and out of the fiber and in connecting fibers are shown in Figure 4.28.

Any phenomenon that causes one amplitude to vary with respect to the other to affect the sum will produce an output. Ordinarily, the laser diode light is linearly polarized in a transverse direction. In common single-mode fibers, there is some variation in refractive index from one transverse plane to another; the amount changes with bending, radial stress, and temperature. This unwanted variable birefringence causes random variable polarization shifts. Rotation of the polarization vectors of two waves being added affects the sum in the same way as time delay phase shifts and produces a spurious output signal.

The best solution to this problem is to deliberately create a birefringence that is stable and large enough to "swamp" any random environmentally induced variations. When polarization-preserving fibers are used, it is essential to launch the light with its polarization vector parallel to one of the principal planes of birefringence – either the "fast" or the "slow" plane. It is also necessary to align these planes when splicing fibers. This considerably increases the complexity of the splicing operation.

4.18.9 Reference Branch Phase Modulator

The phase-locked tracking loop arrangement requires a means for modulating the phase delay of the reference branch to balance the transducer delay. Piezoelectric and magnetic materials are used to strain the fiber. The reference branch fiber is wound under tension (prestressed) around a piezoelectric cylinder fitted with electrodes and so poled that its circumference changes with applied voltage. Minimum cylinder diameter is determined by the fiber bending radius at which fiber transmission losses increase due to "leakage" into the cladding due to the increased ray incidence angles at the core-cladding interface.

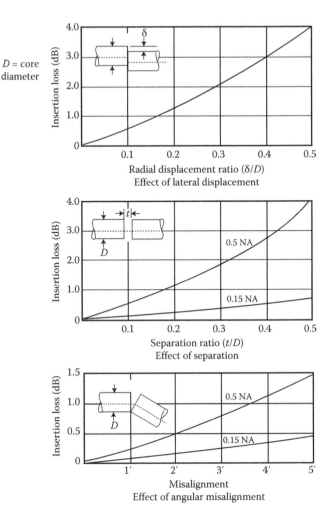

FIGURE 4.28
Effects of lateral displacement, separation, and angular misalignment.

4.18.10 Electronic Circuitry

Optical fiber outputs from the second coupler of the Mach Zehnder interferometer are each coupled to a detector as shown in Figure 4.29—in this case, UDT PHOTO model UDT-455HS. This is a silicon photodiode combined with an op amp in a TO-5 housing. It operates from 9 V positive and negative supply voltages. Gain and bandwidth are adjustable with external feedback components. The outputs are applied to an LM258 connected as a difference amplifier. The dynamic interferometer output signal is taken from the output of the difference amplifier. This output is also low-pass filtered, amplified further by another LM258 and a high-voltage dc amplifier that supplies the control

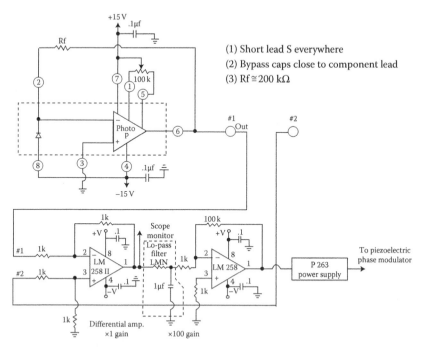

FIGURE 4.29
Interferometer electronics schematic.

voltage to the piezoelectric optical phase modulator in the reference arm of the interferometer. The high-voltage dc amplifier has controls for inserting a continuously variable dc bias if desired. No attempt was made to limit the signal bandwidth in this design, which would improve the SNR if needed.

4.19 Design Example 2: Fiber-Optic-Based Laser Warning Receiver

Military platforms requiring laser warning receiver (LWR) systems include fixed and rotary wing aircraft, ships, and tanks [56]. Timely detection of enemy laser activity, and characterization in terms of the associated weapon system, offers the potential for implementing self-protective action (e.g., evasive maneuver and/or the deployment of a smokescreen) or counterfire.

The lasers against which laser warning may be required are comparatively low in energy and are generally used in association with weapon systems in much the same way as, for example, fire control radars. Thus, detection of illumination by a laser indicates the probable imminent arrival of a munition. Recent developments in LWRs are intended to integrate closely with radar and other threat warning systems, such that multispectral multisensor data clouds provide improved countermeasure implementation and improved vehicle survivability.

4.19.1 System Requirements

The key requirement of an LWR is to provide high probability of intercept for laser threats while maintaining an acceptably low false alarm rate. High probability of intercept implies a wide dynamic range and broadband response to cover all likely threat wavelengths. The LWR must determine enough information about the threat, in terms of bearing and other relevant parameters (e.g., wavelength, pulse width, and pulse repetition frequency) to identify it unambiguously or at least to characterize it sufficiently to enable the timely implementation of appropriate countermeasures. Various fielded systems include unique fiber bundle collection apertures for low ambiguity wide field of view coverage of multiple laser threats, high-resolution optical wavelength analyzers with holographic dispersion gratings, phase delay interpretation for angle of arrival determination, and means of retrieval and identification of laser threat data.

4.19.2 Laser Threats

The most commonly encountered lasers used for range finding and target designation are based on ruby, neodymium, and carbon dioxide systems. CO_2 lasers in the infrared are better able to penetrate rain, haze, and smoke than the older visible and very near infrared. They also have the advantage of being eye safe at low output levels, as are those emitting at 1.54 and 2.06 μm.

Rangefinder pulse widths are typically 10 (Nd:YAG) or 50 ns (CO_2), and pulse repetition rates are between a few and a few tens per second. Output pulse energies are typically ~20 mJ, and beamwidths are ~0.5 mrad. Thus, at a likely engagement range of 1 km, the pulse energy density at the target could be in the region of 10^{-3} J m^{-2}. Considerations of atmospheric transmission, different source-target ranges, and the variety of threat sources lead to the requirement for an LWR with a rather extended dynamic range. This is compounded by the nonuniform illumination across the laser beam (which may be modeled as Gaussian in profile, but where the speckle effect caused by atmospheric turbulence must also be taken into account). Overall, a dynamic range of around 10^6 in optical power will be needed.

4.19.3 Laser Detection

The LWR must detect radiation from pulsed sources operating at discrete wavelengths within the visible to near-infrared spectral range, identify these wavelengths, and accurately measure laser pulse angle of arrival (see Figure 4.30 for a block diagram). Fiber input bundles with collection apertures embedded in the outer surface of a military platform direct incident radiation to a concave holographic grating where the laser radiation is

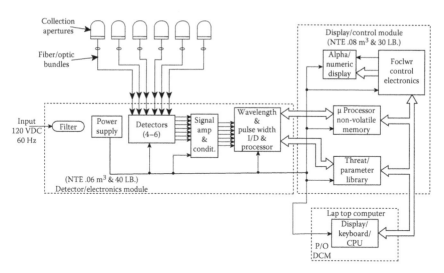

FIGURE 4.30
LWR block diagram.

spectrally dispersed to two linear fiber arrays. Each array will be integrated with discrete detectors. The fiber array located at the grating focal plane will utilize zero order diffraction elements to determine quadrant detection of arrival, while the second array responds to first-order dispersion for wavelength identification. The advantages of this grating arrangement are reduced background clutter as seen by individual detectors and discrimination against false alarms caused by solar glint, chopped sunlight, gun flashes, and strobe lights. The light from broadband sources such as sunlight will be spread over several detectors. Conversely, laser radiation will illuminate one or at most two adjacent detectors. Simple electronic logic will be used to discriminate between laser and nonlaser sources.

4.19.4 System Configuration

The proposed LWR design makes the use of optical fiber bundles to couple laser energy from the outside world to a detector electronics module (see Figure 4.31). Fiber-optic implementation offers many advantages. The removal of detectors and preamplifiers from the skin greatly simplifies the task of eliminating EMI effects. We also have flexibility in locating the detector electronics package, reducing installation and space constraints, and facilitating environmental hardening. Since the input apertures are small, locations for mounting with an unobstructed field of view are more easily found. Small input apertures also result in low optical and radar cross section and minimal aerodynamic disruption. Finally, the elimination of heavy electronic interconnecting cables and multiple detector packages should result in reduced system weight.

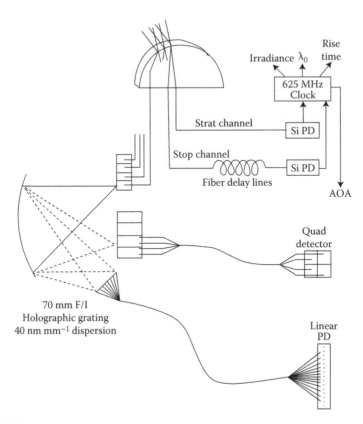

FIGURE 4.31
LWR schematic.

General baseline capabilities might be the following: low false alarm rate, large field of coverage, early warning time (high sensitivity), low visible profile, low radar and optical cross sections, low or zero emissions, ruggedness, and mounting configuration flexibility. Nonimaging LWR systems may have panoramic coverage, but offer only limited directional resolution (typically between 15° and 90°). Interferometric configurations have the ability to test temporal and/or spatial coherence, and potentially to provide both bearing and wavelength data at medium resolution over a reasonable field of view. Difficulties with these systems include the analysis of complex fringe patterns, which may be of poor visibility, particularly under high ambient lighting conditions.

In an imaging LWR, incident laser radiation on the field of view is focused to a spot in an image. The bearing of a detected threat laser can be obtained directly from the location of the spot in the image of the scene. Note that because of the short laser pulse durations, the use of essentially staring rather than scanned sensors is necessary. Such systems do not provide any wavelength output, and temporal information is usually lost with conventional integrating image recording devices such as CCDs. A trade between angular precision and field of view is necessary.

4.19.5 False Alarms and Laser Discrimination

Potential naturally occurring sources of false alarm include direct sunlight, sun glint, and lightning. The sun is by far the strongest broadband source in the visible and near-infrared. Artificial false alarm sources, also broadband, may include searchlights, flares, and munition explosions.

Lasers are often distinguishable from nonlaser sources by their temporal characteristics as well as their narrow linewidths. Although short pulse (tens of nanosecond) laser systems have been discussed thus far, it is conceivable that longer pulsed or quasi-CW lasers could be encountered, and discrimination against typically millisecond duration false alarms could then be difficult. Consideration must also be given to the false alarms, or false bearing indications, generated through reflection of laser radiation off local terrain, clouds, and other vehicles. Discrimination on the basis of spatial coherence analysis may be possible, subject to the complicated effects of atmospheric turbulence and scatter on spatial coherence.

4.19.6 System Summary

Some unique features of the system concept described herein include the following:

- Unique fiber bundle collection apertures for low-ambiguity wide field of view coverage of multiple wavelength laser threats
- High-resolution optical wavelength analyzer using custom holographic dispersion grating and fiber coupler
- Novel optical phase delay coding for precise angle of arrival determination
- Custom algorithms for retrieval and identification of laser threat data

References

1. Mentzer, M. A. 1988. Fiber sensors. *Paper Presented in Proceedings Investigator's Meeting on Semiconductors, Optoelectrics and Magnetic Optic Materials*, Watertown Materials Technology Laboratory, January, Watertown, MA.
2. Ibid.
3. OFC/IOOC 1987 Technical Digest. March 1987. Reprinted in *Lightwave*.
4. Krohn, D. A. 1988. *Fiber Optic Sensors: Fundamentals and Applications*. Research Triangle Park, NC: Instrument Society of America.

5. Davis, C. M. 1982. *Fiberoptic Sensor Technology Handbook*. McLean, VA: Optical Technologies, Inc.
6. Dakin, J. and B. Culshaw, eds. 1988. *Optical Fiber Sensors: Principals and Components*. Boston, MA: Artech House.
7. Hocker, G. B. 1979. Fiber-optic sensing of pressure and temperature. *Appl. Opt.* 18(9):1445–1448.
8. White, R. M. August 1983. Magnetic disks: storage densities on the rise. *IEEE Spectrum*:20(8):32.
9. Mitchell, G. L., J. E. Lenz, and C. D. Anderson. 1982. Final report for contract F086635-81-C-0311. Honeywell Systems and Research Center, July, Minneapolis, MN.
10. Spano, M. L., K. B. Hathaway, and H. T. Savage. 1998. *J. Appl. Phys.* 53:2667.
11. Ennen, H., G. Pomrenke. A. Axemann, K. Eisel, W. Haydl, and J. Schneider. 1985. *Appl. Phys. Lett.* 46:381.
12. Uttman, D. and B. Culshaw. 1985. Semiconductor lasers in advance interferometric optical fiber sensors. *IEEE Proc. J.* 132(3):184.
13. Lau, K. W., C. Harder, and A. Yariv. 1984. Longitudinal mode spectrum of semiconductor lasers under high-speed modulation. *IEEE J. Quant. Electron.* QE-20(1):71–79.
14. Mentzer, M. A. 1990. *Principles of Optical Circuit Engineering*. New York: Marcel Dekker.
15. Slonecker, M. H. April 1982. Paper WBB7 in *OFC-82 Digest of Technical Papers*. Washington, DC: Optical Society of America.
16. Jeunhomme, L. B. 1983. *Single-Mode Fiber Optics*. New York: Marcel Dekker, 172 ff.
17. Cancellieri, G., P. Fentini, and U. Pesciarelli. 1985. Effects of joints on single-mode polarization optical fiber links. *Appl. Opt.* 24(7):964 ff.
18. Lau, K. W., C. Harder, and A. Yariv. 1983. Longitudinal mode spectrum of GaAs injection lasers. *Appl. Phys. Lett.* 43(7):619–621.
19. Uttman, D. and B. Culshaw.. Semiconductor lasers. *IEEE*.
20. Carr, W. N. 1965. Characteristics of a GaAS spontaneous infrared source with 40 percent efficiency. *IEEE Trans. Electron Devices* ED-12:531.
21. Sze, S. M. 1981. *Physics of Semiconductor Devices*, 2nd edn. New York: John Wiley & Sons, pp. 683–686.
22. Lasher, G. J. 1963. Spontaneous and stimulated recombination radiation in semiconductors. *IBM J.* 7:58.
23. Sze, S. M. 1981. *Physics of Semiconductor Devices*. New York: Wiley.
24. Ogawa, K. 1982. Analysis of mode partition noise in laser transmission systems. *IEEE J. Quant. Electron.* QE-18(5):849–855.
25. Lau, K. W., C. Harder, and A. Yariv. Longitudinal mode spectrum. *IEEE*.
26. Ibid.
27. Nakamura, K. A., N. Chinone, R. Ito, and J. Umeda. 1978. Longitudinal mode behavior of mode-stabilized $Al_xGa_{1-x}As$ injection lasers. *J. Appl. Phys.* 49:4644–4648.
28. Lang, R. and K. Kobayashi. 1980. External optical feedback effects on semiconductor injection laser properties. *IEEE J. Quant. Electron.* QE-16:347–355.
29. Sakakibara, Y., K. Furuva, K. Utaka, and Y. Suematsu. 1980. Single mode oscillation under high speed direct modulation in GaInAsP/InP integrated twin-guide lasers with first order distributed Bragg reflectors. *Electron. Lett.* 16:456–458.

Utaka, K., I. Kobayashi, and Y. Suematsu. 1981. Lasting characteristics of 1.5–1.6 μm GaInAsP/InP integrated twin-guide lasers with first order distributed Bragg reflectors. *IEEE J. Quant. Electron.* QE-17:651–568.

30. Lee, T. P., C. A. Burrus, P. L. Liu, and A. G. Dentai. 1982. High efficiency short-cavity InGaAsP lasers with one high reflectivity mirror. *Electron. Lett.* 18:805–807.
    Lee, T. P., C. A. Burrus, R. A. Linke, and R. J. Nelson. 1983. Short cavity single frequency InGaAsP buried heterostructure lasers. *Electron. Lett.* 18:82–84.

31. Ebeling, K. J., L. A. Coldren, B. I. Miller, and J. A. Rentschler. 1983. Single mode operation of coupled cavity GaInAsp/InP semiconductor lasers. *Appl. Phys. Lett.* 42:6–8.

32. Ettenberg, M. September 1987. Laser diode systems and devices. *IEEE Circuits Devices Mag.*: 22–26.

33. Hunsberger, R. G. 1985. *Integrated Optics Theory and Technology*, 2nd edn. New York: Springer-Verlag.

34. Lau, K. W., C. Harder, and A. Yariv. Longitudinal mode spectrum. *IEEE*.

35. Ibid.

36. Henry, C. H. and R. F. Kazarinov. 1986. Instability of semiconductor lasers due to optical feedback from distant reflectors. *IEEE J. Quant. Electron.* QE-22:294–301.

37. Ibid.

38. Lang, R. and K. Kobayashi. External optical feedback effects. *IEEE J. Quant. Elect.* QE. 16(3):347–355.
    Acket, G. A., D. Lenstra, A. J. denBoef, and B. H. Verbeek. 1984. The influence of feedback intensity on longitudinal mode properties and optical noise in index-graded semiconductor devices. *IEEE J. Quant. Electron.* QE-20:1163–1169.
    Osmundsen, J. H., B. Tromborg, and H. Olesen. 1984. Experimental investigation of stability properties for a semiconductor laser with optical feedback. *Electron. Lett.* 19:1068–1070.

39. Patzak, E., H. Olesen, A. Sagimura, S. Sato, and T. Mukai. 1983. Spectral linewidth reduction in semiconductor lasers by an external cavity with weak optical feedback. *Electron. Lett.* 19:938–949.
    Kikuchi, K. and T. Okoshi. 1982. Simple formula giving spectrum-narrowing ratio of semiconductor laser output obtained by optical feedback. *Electron. Lett.* 18:10–12.
    Agrawal, G. P. 1984. Line narrowing in a single-mode injection laser due to external optical feedback. *IEEE J. Quant. Electron.* QE-20:468–471.

40. Miles, R. O., A. Danbridge, A. B. Tveten, H. F. Taylor, and T. G. Giallorenzi. 1980. Feedback-induced line broadening in CW channel substrate planar laser diodes. *Appl. Phys. Lett.* 37:990–992.
    Goldberg, L., H. F. Taylor, A. Dandridge, J. F. Weller, and R. O. Miles. 1980. Spectral characteristics of semiconductor lasers with optical feedback. *IEEE Trans. Microwave Theory Tech.* MTT-30:401–410.
    Lenstra, D., B. H. Verbeek, and A. J. denBoef. 1985. Coherence collapse in single-mode semiconductor lasers due to optical feedback. *IEEE J. Quant. Electron.* QE- 21, 674–679.

41. Risch, C. and C. Voumard. 1977. Self-pulsation in the output intensity spectrum of GaAs-AlGaAs CW diode lasers coupled to a frequency selective external optical cavity. *J. Appl. Phys.* 48:2083–2085.

Morikawa, T., Y. Mitsuhashi, J. Shimoda, and Y. Kojima. 1976. Return-beam-induced oscillations in self-coupled semiconductor lasers. *J. Electron. Lett.* 12:435–436.

Fujiwara, M., K. Kubota, and R. Lang. 1981. Low frequency intensity fluctuation in laser diodes with external optical feedback. *Appl. Phys. Lett.* 38:217–220.

42. Henry, C. and R. Kazarinov. 2002. Instability of semiconductor lasers. *IEEE J. Quant. Electron.* Q E19:1391–1397.

43. Mentzer, M. A. Fiber sensors. Lectures at Watertown Arsenal. Boston. 1988.

44. Ennen, H., G. Pomrenke, A. Axmann, K. Eisele, W. Haydl, J. Schneider. 1985. *Appl. Phys. Lett.* 46:381.

45. Uttman, D. and B. Culshaw. Semiconductor lasers. *IEEE*.

46. Ennen, H. et al. *Appl. Phys. Lett.*

47. Tai, S., K. Kyuma, and T. Nakayama. 1985. Simultaneous stabilization of spectral linewidth and oscillation frequency of an external cavity laser diode by fiber optic ring resonators. *Proceedings of IOOC-ECOC'85*, pp. 833–836.

    Saito, S. 1982. *IEEE J. Quant. Electron.* QE-18:961.

    Goldberg, L. 1982. Injection locking GaA/As lasers. *IEEE J. Quant. Electron.* QE-18:543–555.

48. Tai, Kyuma, and Nakayama. 1987. Simultaneous stabilization of spectral linewidth. *Elect. Comm. in Japan.* 70(6):19–26.

49. Stokes, L. F. 1982. Ring resonators. *Opt. Lett.* 7:288.

50. Tsang, W. T., and R. A. Logan. 1986. Observation of enhanced single longitudinal mode operation in 1.5-μm GaInAsP erbium-doped injection lasers. *Appl. Phys. Lett.* 49(25):1686–1688.

51. Dieke, G. H. 1981. *Spectra and Energy Levels of Rare-Earth Ions in Crystals.* New York: Wiley Interscience.

    Kasatkin, V. A., F. P. Kesamanly, and B. E. Samourukov. 1981. *Sov. Phys. Semicond.* 15:352.

    Zakharenkov, L. F., V. A. Kasatkin, F. P. Kesamanly, B. E. Samourukov, and M. A. Sokolova. 1981. *Sov. Phys. Semicond.* 15:946.

    Ennen, H., U. Kauffaman, G. Pomrenke, J. Schneider, J. Windschief, and A. Axmann. 1983. *J. Cryst. Growth* 64:165.

    Ennen, H. and J. Schneider. 1985. *J. Electron. Mater.* 14A:115.

    Dmitriev, A. G., L. F. Zakharenhov, V. A. Kasatkin, V. F. Masterov, and B. E. Samourikov. 1983. *Sov. Phys. Semicond.* 17:1201.

    Haydl, W. H., H. D. Muller, H. Ennen, W. Koerber, and K. W. Benz. 1985. *Appl. Phys. Lett.* 46:870.

    Pomrenke, G. S., H. Ennen, and W. Haydl. 1986. *J. Appl. Phys.* 59:601.

    Ennen, H., J. Schneider, G. Pomrenke, and A. Axeman. 1983. *Appl. Phys. Lett.* 43:943.

52. Kirby, P. A. 1981. Semiconductor laser sources for optical communications. *Radio Electron. Eng.* 51:362–367.

53. Sze, S. M. *Physics of Semiconductor Devices*, pp. 683–686.

54. Mentzer, M. A. *Fiber Sensors.* Watertown Arsenal.

55. Giallorenzi, T., J. Bucaro, A. Dandridge, G. Sigel, J. Cole, and S. Rasleigh. 1982. Optical fiber sensor technology. *IEEE J. Quant. Electron.* QE-8(4):626–665.

56. Mentzer, M. A. *Fiber Sensors.* Watertown Arsenal. Meeting on optical switching, 1986.

5

Integrated Optics

5.1 Planar Optical Waveguide Theory

We first discuss the case of the planar waveguide in terms of the carrier compensation process. If the waveguide core layer is assumed homogeneous, then the amount of free-carrier compensation necessary for waveguiding, that is, at least for the fundamental transverse electric mode (TE$_0$) to propagate, can be calculated. This cutoff condition for TE$_0$ to propagate is [1]

$$\frac{2\pi d}{\lambda_0}\sqrt{n_1^2 - n_2^2} \geq \arctan\frac{n_2^2 - n_3^2}{n_1^2 - n_3^2} \tag{5.1}$$

Since $n_2 \gg n_3$ and $n_1 \geq n_2$, the argument of the inverse tangent is much greater than unity but is not always large enough in the cases considered for arctan(x) to be approximated by $\pi/2$. Therefore, the simple mathematical identify arctan(x) = $\pi/2$—arctan($1/x$) allows an accurate Taylor series approximation to be made. Note that

$$\arctan(m) = m - \frac{m^3}{3} + \frac{m^5}{5} \mp \cdots \cong m \tag{5.2}$$

and that

$$n_1^2 - n_2^2 \cong 2n_2\Delta n \tag{5.3}$$

results in the following condition, which is the minimum index change necessary for waveguiding:

$$n_1 - n_2 = \Delta n \geq \frac{\pi^2}{8n_2}\left(\frac{1}{2\pi d/\lambda_0 + 1/(n_2^2 - n_3^2)}\right)^2 \tag{5.4}$$

Expressing Δn in terms of the free-carrier concentrations in both the core and substrate regions gives the carrier compensation necessary for waveguiding to occur:

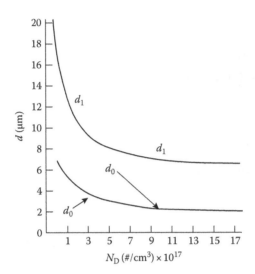

FIGURE 5.1
Cutoff thickness versus substrate doping for $\lambda_0 = 1.3\,\mu m$.

$$N_3 - N_1 \geq \frac{\varepsilon_0 m^+ \omega^2 \pi^2}{4e^2} \left(\frac{1}{2\pi d / \lambda_0 + 1/(n_2^2 - n_3^2)} \right)^2 \tag{5.5}$$

In general, the carrier compensation necessary for guiding the TE_1 (1-th order) mode is given by the expression

$$N_3 - N_1 \geq \frac{\varepsilon_0 m^+ \omega^2 \pi^2}{4e^2} \left(1 + \frac{1}{2} \right)^2 \left(\frac{1}{2\pi d / \lambda_0 + 1/(n_2^2 - n_3^2)} \right)^2 \tag{5.6}$$

Figures 5.1 through 5.3 illustrate representative cutoff thickness versus substrate doping curves for three wavelengths of interest. The assumption is made that essentially complete carrier compensation will occur in the waveguide fabrication process.

5.2 Comparison of "Exact" and Numerical Channel Waveguide Theories

A number of workers have analyzed lossless dielectric waveguide structures including Marcatili, Goell, Schlosser, Bartling, and Shaw [2]. Goell represented the radial variation of the longitudinal electric and magnetic fields of

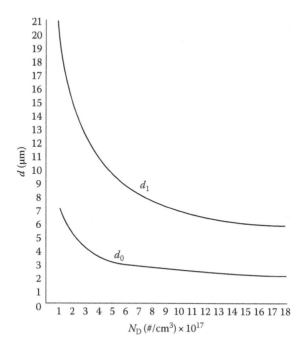

FIGURE 5.2
Cutoff thickness versus substrate doping for $\lambda_0 = 3.4\,\mu m$.

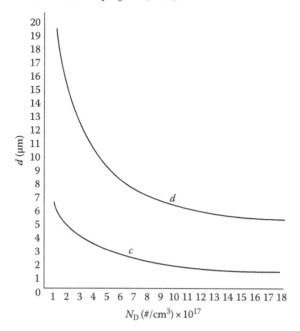

FIGURE 5.3
Cutoff thickness versus substrate doping for $\lambda_0 = 10\,\mu m$.

the circular dielectric waveguide modes by a sum of modified Bessel functions inside and outside the waveguide core. Matching the fields along the perimeter of the core provided computer solutions. Schlosser's analysis was for the case of a dielectric rod in air and applied to index of refraction differences more applicable to microwaves. Bartling solved the problem in terms of a Green's dyadic but did not calculate specific cases.

After some simplifying assumptions, Marcatili solved Maxwell's equations in closed form for rectangular cross section dielectric waveguides of various aspect ratios. Proper selection of the index of refraction results in a waveguide supporting only the fundamental modes in a family of two types of hybrid transverse electric magnetic (TEM) modes polarized at right angles.

Marcatili's analysis provides analytical results in a relatively simple form and is preferred since the calculations required are simpler. Goell suggests that the analysis may not provide the accuracy desired for the lowest order mode, however [3], as shown in Figure 5.4. Figure 5.4 provides solutions from Marcatili's transcendental and closed form equations and Goell's circular harmonic solutions for an aspect ratio (channel width to height ratio) of 2, with an index difference in the channel greater than 1.05 times that of the surrounding dielectric. The two methods indicate values of a normalized propagation constant (in terms of the transverse and waveguide propagation constants)

$$p^2 = \frac{k_z^2 - k_4^2}{k_1^2 - k_4^2} \cong \left(\frac{k_z}{k_4} - 1\right)\left(\frac{n_1}{n_4} - 1\right) \tag{5.7}$$

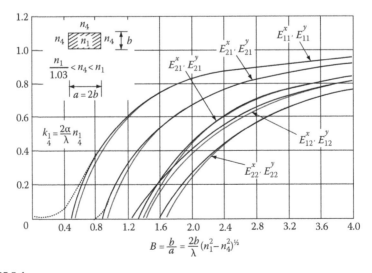

FIGURE 5.4
Propagation constant for several modes of rectangular dielectric waveguide. Transcendental solution ⋯⋯; closed form solutions ——; Goell's computer solution.

which are within a few percent for $p^2 > 0.5$, although Goell indicates a "fanning out" to the left on the curves for small p^2.

An analysis performed by Bradley and Kellner [4] was used to verify the accuracy of Marcatili's approach as applied to carrier compensated waveguides. The numerical analysis program uses cubic spline functions in a finite element analysis. The comparison indicates essentially perfect agreement with the modal structure and cutoff conditions predicted by the Marcatili analysis (excepting very small p^2 values), so that the ease afforded by the use of Marcatili's approach does not significantly compromise the accuracy or choice of fabrication parameters.

5.3 Modes of the Channel Waveguide

The basic channel structure is shown in Figure 5.5. In many cases of interest, $n_3 = n_4 = n_5 < n_1$ and $n_2 = n_1$; however, the more general case will be discussed first. A simplification at this point enables a closed form solution. The key assumption is that the modes are well above cutoff, so that the field decays exponentially in regions 2, 3, 4, and 5, with most of the power traveling in region 1. Very little power travels in the shaded regions of Figure 5.5 so that, by not matching the fields along the edges of the shaded areas, we are not significantly affecting the calculation of fields in region 1.

The modes are essentially of the TEM type and can be grouped into two families, E_{pq}^x and E_{pq}^y, with p and q indicating the number of extrema in the field distributions in the x- and y-directions, respectively. The transverse field components of the E_{pq}^x modes are E_x and H_y, while those of the E_{pq}^y modes

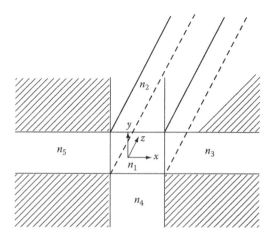

FIGURE 5.5
Ion-implanted channel waveguide structure.

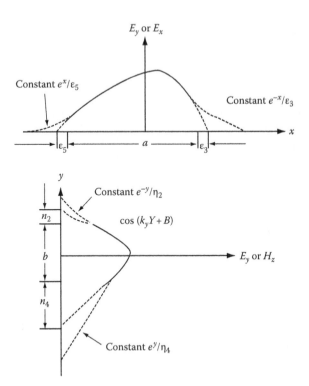

FIGURE 5.6
Field distribution of the fundamental mode$_{11}$ for a guide immersed in several dielectrics.

are E_y and H_x. Figure 5.6 shows the fundamental mode E_{11}, characterized by extinction coefficients n_2, e_3, n_4, and e_5 in the regions where the mode is exponential, and by propagation constants k_x and k_y in region 1. In cases where $n_3 = n_5$, the fields will be symmetric with respect to the $x = 0$ plane.

Quantitative expressions for k_x, k_y, n, and e are now determined.

Assuming that the plane wavelets that make a mode impinge on the interfaces at grazing angles, that is, $1/n_1(n_1 - n_y) \ll 1$, then k, e_5^3, and h_4^2 are given by

$$k_x = \frac{p\pi}{a}\left(1 + \frac{A_3 + A_5}{\pi a}\right)^{-1} \tag{5.8}$$

$$k_y = \frac{q\pi}{b}\left(\frac{1 + n_2^2 A_2 + n_4^2 A_4}{\pi n_1^2 b}\right)^{-1} \tag{5.9}$$

$$k_z = \left[k_1^1 - \left(\frac{\pi p}{a}\right)^2\left(1 + \frac{A_3 + A_5}{\pi a}\right)^{-2} - \left(\frac{\pi q}{b}\right)^2\left(1 + \frac{n_2^2 A_3 + n_4^2 A_4}{\pi n_1^2 b}\right)^{-2}\right]^{\frac{1}{2}} \tag{5.10}$$

$$e_5^3 = \frac{A_5^3}{\pi}\left[1-\left(\frac{pA_5^3}{a}\frac{1}{1+A_3+A_5/\pi a}\right)^2\right]^{-\frac{1}{2}} \quad (5.11)$$

$$n_4^2 = \frac{A_4^2}{\pi}\left[1-\left(\frac{qA_4^2}{b}\frac{1}{1+n_2^2A_2+n_4^2A_4/\pi n_1^2 b}\right)^2\right]^{-\frac{1}{2}} \quad (5.12)$$

$$A_{2,3,4,5} = \frac{\pi}{\left(k_1^2-k_{2,3,4,5}^2\right)^{\frac{1}{2}}} = \frac{\lambda_0}{2\left(n_1^2-n_{2,3,4,5}^2\right)^{\frac{1}{2}}} \quad (5.13)$$

The E_{pq}^y modes are polarized such that E_y is the only significant component of electric field with E_x and E_z negligibly small. In the case of the E_{pq}^x modes, E_x is the only significant electric field component.

A computer code was written to calculate the modes of channel waveguides using Marcatili's equations. The program provides the e's, n's, and n_{eff}'s$=k_z$'s for any particular channel guide design. The goal of the design process was to determine the fabrication parameters that would produce a guide with the least amount of propagation loss possible and with only single mode propagation. The key factor in the determination of these optimum parameters is the minimization of the optical power guided in the mode tails residing in the substrate, that is, n_3, n_4, n_5 region. Calculations of waveguide cutoff curves and losses for any aspect ratio can be performed for any propagation wavelength. Figures 5.7 through 5.9 are the propagation cutoff curves at 1.3, 3.4, and 10.6 µm wavelengths for an aspect ratio of 1. These curves, along with the accompanying loss versus N_D curves, provide the basis for the selection of N_D, t, and aspect ratio at each wavelength for channel waveguides. In addition, Figures 5.10 through 5.12 indicate the specific range or distance on the cutoff curve x-axis, over which the E_{11}^x and E_{11}^y modes are guided but the E_{21}^x and E_{21}^y modes are not guided, versus N_D and aspect ratio. This thickness tolerance is important since it decreases, for example, with increasing aspect ratio. Since a thickness is generally chosen somewhere near the center of this range to prevent any fabrication error causing either multimode or no mode guiding, this range is an important design consideration.

5.4 Directional Couplers

The dual channel directional coupler consists of parallel channel waveguides spaced so that synchronous coherent coupling between the overlapping

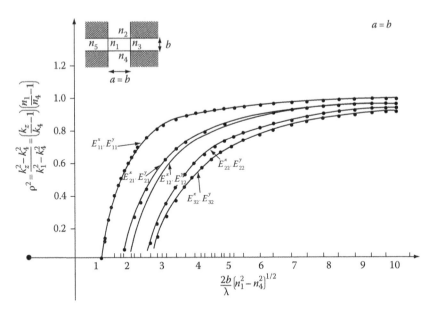

FIGURE 5.7
Propagation cutoff curve for $\lambda_0 = 1.3\,\mu m$.

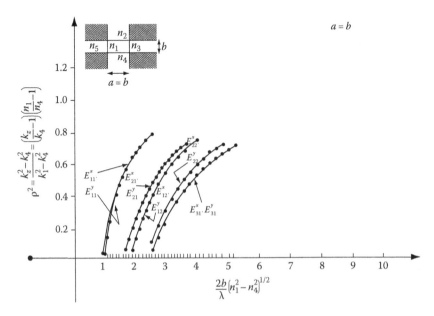

FIGURE 5.8
Propagation cutoff curve for $\lambda_0 = 3.4\,\mu m$.

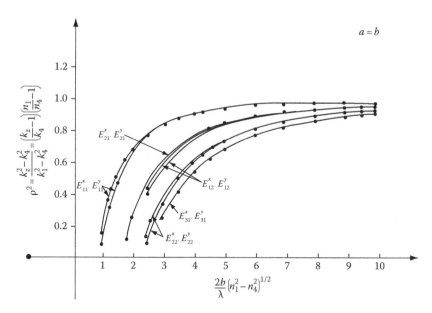

FIGURE 5.9
Propagation cutoff curve for $\lambda = 10.6\,\mu m$.

evanescent mode tails occurs. Figure 5.13 illustrates such a device. The electrical properties and field penetrations e_{35} and n_{24} for the parallel guides shown in Figure 5.13 are the same as those of the isolated channel. Such a parallel channel directional coupler can transmit both the $E_{pq}{}^x$ and $E_{pq}{}^y$ modes.

Following the analysis of Marcatili [5], the coupling coefficient K between the two guides and the length L (along the z-direction) necessary for complete transfer of power from one to the other is given by

$$-iK = \left(\frac{\pi}{2L}\right) = 2 * \left(\frac{Kx^2 * e_5 e^{-c/e_5}}{k_z a * (1 + k_x^2 e_5^2)}\right) \tag{5.14}$$

A more-accurate expression for the coupling was published by Kuznetsov [6]. This work indicated that the Marcatili analysis overestimates the coupling coefficient, especially in the weaker-guiding cases, by as much as a factor of 2. The design program used for channel guides was also written to calculate the coupling length L for any separation C, for identical guides, for the X-polarized electrical field. Both the Marcatili and the Kuznetsov values of coupling length have been determined as a direct basis for comparison.

Following are plots of the predicted coupling lengths versus channel separations for several representative channel waveguide designs (Figure 5.14a

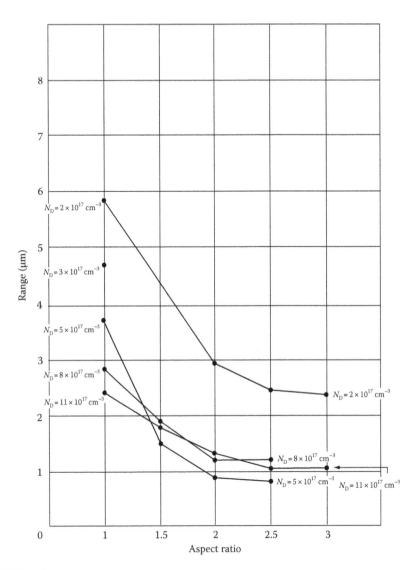

FIGURE 5.10
$\lambda_0 = 1.3\,\mu m$—Maximum range of thickness for guiding only the $E_{11}^x \cdot E_{11}^y$ modes.

through e). Accurate control of the guide separation is very important since the coupling length L decreases exponentially with the ratio c/e_5, where e_5 is the transverse mode extinction coefficient. In fact, for separations $C < e_5$ there will be a significant perturbation upon the optical mode field profile by each channel upon the other. Predictions of the coupling length for separations less than e_5, therefore, would require a more extensive analysis of the perturbation effect.

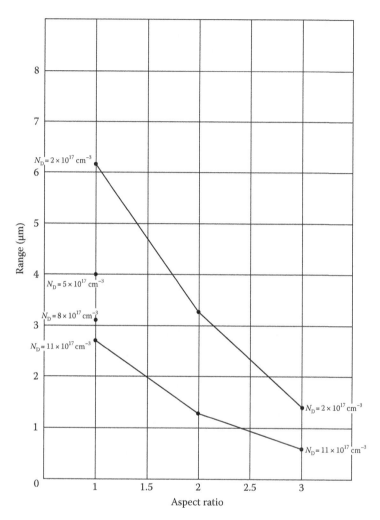

FIGURE 5.11

$\lambda_0 = 3.4\,\mu m$—Maximum range of thickness for guiding only the $E_{11}^x \cdot E_{11}^y$ modes.

5.5 Key Considerations in the Specifications of an Optical Circuit

5.5.1 Introduction

A number of key interrelated issues need to be addressed in order to properly specify an integrated optical circuit. These considerations include operational wavelength, optical throughput losses, material growth, electronic

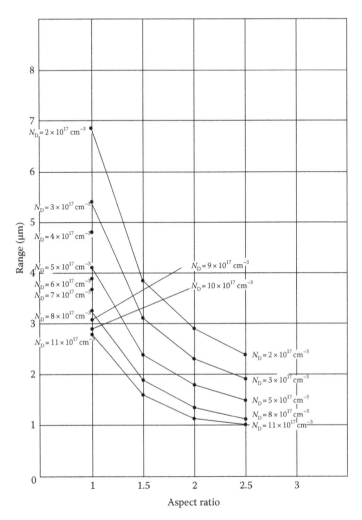

FIGURE 5.12
$\lambda_0 = 10.6\,\mu m$—Maximum range of thickness for guiding only the $E_{11}^x \cdot E_{11}^y$ modes.

and microwave compatibility, and integratability. Much of the work in the field of optical circuit engineering has involved optimization of a particular performance aspect. The next step is to realize fully optimized integrated optical circuit systems.

Of all the considerations involved in the selection of substrate material, optical loss is often the overriding factor. Optical loss is generally due to artifacts of fabrication such as surface roughness, interface strain and defects, nonideal or non-abrupt dielectric discontinuity, and unwanted interstitial defect trapping and scattering centers. Aside from optical absorption occurring via free-carrier absorption of power residing in heavily doped regions

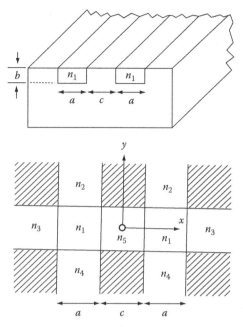

FIGURE 5.13
Parallel channel directional coupler.

of the optical mode profile, it is the interband or band-edge absorption which will dominate the optical power budget. Purity and control of the fabrication environment provide both low-loss material (e.g., minimize unwanted tailing of the absorption band edge), and optimum surface and interference quality for both efficient electro-optic interaction and maximum optical throughput.

Intrinsic semiconductor loss is primarily due to free-carrier absorption, so that selection of the 0.86 µm wavelength is not a disadvantage in terms of loss relative to 1.3 µm provided the band edge of the optical waveguide relative to the light source is properly chosen. Losses due to interband absorption are a serious consideration at 0.86 µm, but with enhanced material growth and fabrication capability, considerably improved quality materials (i.e., vastly improved control of defect concentration) will be more readily available.

The larger amount of waveguiding loss at 0.86 µm can be compensated for by several approaches. First, at 0.86 a GaAlAs laser diode can be integrated with other optical components on the same chip to reduce coupling loss. The required technology to do this at 1.3 µm is considerably lacking. Fabrication of microwave transistor monolithic integrated circuits and digital integrated circuits is much farther advanced for GaAs compared to InP. This alone might suggest the choice of GaAs as a substrate material. Optical amplifiers at 0.86 µm can be built with better than 20 dB gain [7]; therefore in terms of system loss, the addition of this one additional component on the chip is advantageous. Furthermore, since the structure required for the optical amplifier is identical to the laser diode structure, the fabrication complexity will not be significantly increased.

Numerous current and downstream applications for 0.86 µm integrated optics technology provide substantial justification for its continued development. A powerful motivation for working in the 0.86 µm region is the blossoming area of nonlinear optics, or exitonic absorption phenomena, which are expected to permit the realization of modulation and switching speeds on the order of 100 fs (10^{-13} s). Optical systems with this capability will find application in optical signal processing for digital computation and for total information wide-band processing as required in the multi-information

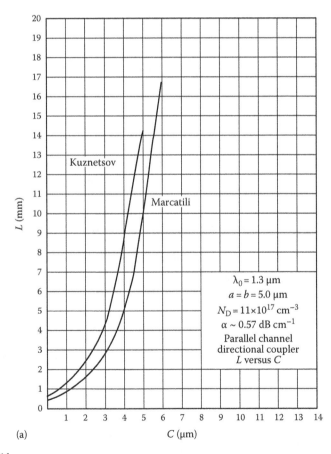

(a)

FIGURE 5.14

(a) $\lambda_0 = 1.3 \mu m$; $a = b = 5.0 \mu m$; $N_D = 11 \times 10^{17}$ cm^{-3} parallel direction channel directional coupler L versus C.

coded channel scenario of the modern battlefield, and in the complexity of the international and satellite audio and visual communications [8]. These structures are extremely important from the stand point of basic science because their submicron geometry necessitates altering the traditional, mathematical, and physical analysis.

The material quality and critical fabrication tolerances afforded through MOCVD and the MBE permit fabrication of single crystals with nanometer dimensions and ultimately single monatomic layers. This gives rise to previously unexplored conditions such as ellipsoidal symmetry of the exciton and layer thicknesses less than the mean free path of the electron, and hence effects such as exciton stability at room temperature and ballistic tunneling of electrons and hot electrons. These effects will ultimately be implemented in a guided-wave integrated format, affording improved optical system performance.

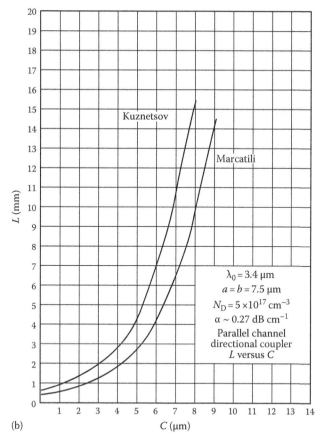

(b)

FIGURE 5.14 (continued)
(b) $\lambda_0 = 3.4\,\mu m$; $a = b = 7.5\,\mu m$; $N_D = 5 \times 10^{17}$ cm^{-3}; $\alpha \sim 0.27$ db cm^{-1} parallel direction channel directional coupler L versus C.

(continued)

5.5.2 Waveguide Building Block and Wavelength Selection

Various fundamental waveguide structures have been investigated for use with GaAs substrates (LiNbO and other material systems will be discussed in a later section). These structures will be reviewed in this section along with the factors leading to the final preferred choice. Several structures are made without the use of multiple materials. Ion implantation has been shown to produce the necessary change in the refractive index of the substrate material to induce waveguiding [9]; however, free-carrier absorption loss due to the heavily doped substrates results in losses on the order of 1 dB cm^{-1} [10].

Another alternative is the use of deposited/etched thin-film dielectric waveguides [11] on the surface of the GaAs wafer. This is a fairly attractive option because a waveguide material such as SiO_2 or Ta_2O_2 would not exhibit

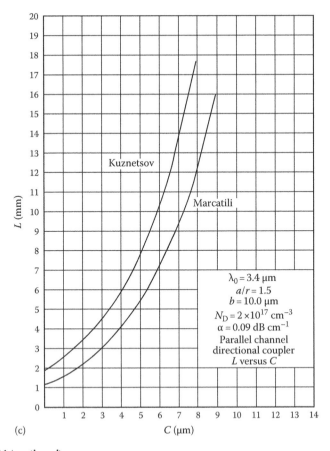

(c)

FIGURE 5.14 (continued)
(c) $\lambda_0 = 1.3\,\mu m$; $a/r = 1.5$; $b = 10.0\,\mu m$; $N_D = 2 \times 10^{17}$ cm^{-3}; $\alpha = 0.09$ db cm^{-1} parallel direction channel directional coupler L versus C.

either interband absorption or free-carrier absorption. It might, therefore, be possible to obtain less than 1.0 dB cm⁻¹ loss over the entire wavelength range of 0.82–1.3 μm; however it is very difficult to couple light from a device such as a laser or modulator in the GaAs wafer into the waveguide without considerable coupling loss because the index of refraction of GaAs is so much larger than that of the dielectric. The three exposed surfaces of the dielectric waveguide are also subject to damage and degradation due to physical abrasion and atmospheric contaminants. To act as a waveguiding region the deposited film would need to be isolated from the GaAs substrate by another film having a lower refractive index than the waveguiding film. In the past it has been difficult to deposit such a two-layer structure and achieve low waveguiding loss. There is also the possibility of the waveguide lifting from the GaAs surface because of mismatch in the thermal expansion coefficients.

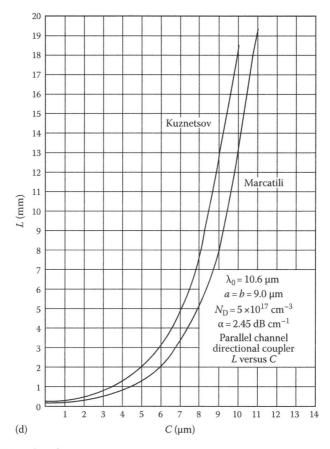

(d)

FIGURE 5.14 (continued)
(d) $\lambda_0 = 10.6\,\mu m$; $a = b = 9.0\,\mu m$; $N_D = 5 \times 10^{17}$ cm⁻³; $\alpha = 2.45$ db cm⁻¹ parallel direction channel directional coupler L versus C.

(continued)

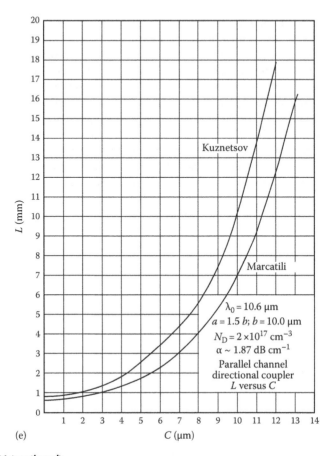

(e)

FIGURE 5.14 (continued)
(e) $\lambda_0 = 10.6\,\mu m$; $a = 1.5b$; $b = 10.0\,\mu m$; $N_D = 2 \times 10^{17}$ cm^{-3}; $\alpha \sim 1.87$ db cm^{-1} parallel direction channel directional coupler L versus C.

Other structures can be formed by growing epitaxial AlGaAs layers on the substrate. These layers yield a planar waveguide, providing mode confinement in the vertical direction. Some form of etching or ion milling is then used to define a structure that also provides confinement in the lateral direction. Three such structures are the raised rib, the buried encapsulated, and the buried strip-loaded waveguides. Schematic diagrams of these structures are shown in Figure 5.15.

The first two structures have some obvious disadvantages. The raised rib design has most of the confined mode present in the rib. This is subject to a great deal of scattering loss from the etched side walls of the guide. Because the guide is on the surface, it is also subject to damage and contamination. The encapsulated design avoids this second pitfall but relies on more complicated lithographic techniques and material growth. This latter aspect

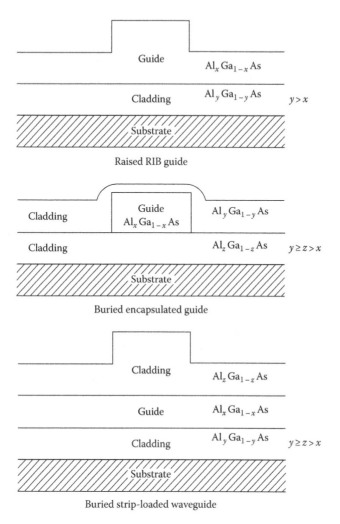

FIGURE 5.15
Fundamental AlGaAs waveguides.

presents significant difficulty when considering electronic and full optical integration in conjunction with the waveguides.

For the above reasons the buried strip-loaded structure may be the optimum approach. This structure has better vertical confinement which keeps the guided energy away from the surface and minimizes scattering losses. The fact that the etched boundaries are located within the cladding layer also reduces their effect on scattering losses, particularly in comparison to the raised rib guide. The additional layering provides for the incorporation of fabrication features such as a stop etch layer. In addition this structure gives better p–n junction characteristics when an electro-optic modulator is being fabricated and provides greater design freedom for the problem of contacting. One detrimental feature of this structure is that it has less lateral confinement, making it more lossy for waveguide bends if not very carefully designed. Numerical examples in another section illustrate this point.

The fundamental constraint on the operational wavelength of this waveguide is normally that the light be confined in a single mode. With a GaAs substrate, the clear choice for the waveguide structure is the AlGaAs/GaAs system. The AlGaAs system will guide light over a range of wavelengths from less than 0.8 μm to greater than 10.6 μm. For AlGaAs waveguides, the wavelengths that have received the most attention are 0.85 and 1.3 μm. Since wavelengths longer than about 0.87 μm are not possible with an integrated source in the AlGaAs material, the final constraint becomes the wavelength of the source. If operation speed is to be in excess of 3 Gb s^{-1}, performance can be improved by integrating the lasers and electronic drive circuitry with the rest of the optical components.

The one performance parameter that probably will not exceed that which can be done at 1.3 μm is optical waveguide loss. Losses in AlGaAs waveguides are typically 1 dB cm^{-1} while losses at 0.85 μm are often 2–4 dB cm^{-1} due to band-edge material loss. As available material quality improves the optical losses will improve. One must consider not only waveguide absorption and scattering loss but also laser to waveguide coupling loss and the advantages to be gained from monolithic integration. Without anti-reflection coating applied to the facets of the waveguide, it is difficult to achieve better than 50% coupling efficiency into an AlGaAs waveguide since reflection loss itself is 30%. End fire or butt coupling losses will be dramatically reduced when the laser is integrated on the substrate. The AlGaAs multilayer material system is much more developed than the InGaAaP so that one would expect lower values of waveguide loss with the former when lasers are monolithically integrated. Although use of the InGaAsP lasers at 1.3 μm wavelength as sources for waveguides formed in AlGaAs can achieve lower values of loss, one gives up the potential for monolithic integration of lasers with integrated optical devices. Another method that can be used to compensate for optical loss is to integrate an optical amplifier on the substrate. Optical amplifiers exhibited gains in excess of 20 dB [12]. Although this demonstration was at a wavelength of 1.5 μm, it could also be achieved in AlGaAs waveguides at 0.85 μm.

The 1.3 μm lasers consist of InGaAsP layers grown on InP substrates. This wavelength is a good choice for specialized long haul communications (distances longer than 10 km); however, because GaAs is transparent at this wavelength, the fabrication of lasers and detectors on the same substrate with GaAs electronics would be extremely difficult. In addition, there is a large lattice mismatch between InP and GaAs.

Higher reliability is achieved by reducing the number of components via larger scale integration. Since sources and detectors can be readily integrated on GaAs substrates the number of hybrid components can be reduced. The least reliable part of any electronic or optoelectronic package is often the wire bonds. With a single monolithic integrated optic/optoelectronic chip the number of wire bonds decreases and reliability improves. This can also be translated to lower cost since the entire process will be subjected to well-calibrated batch fabrication processes.

An additional advantage of increased flexibility exists at the $0.85\,\mu m$ wavelength. There are large research and development efforts in fabricating optical logic devices using AlGaAs/GaAs multiple quantum well structures. These devices rely on a resonant excitonic absorption and must be operated near the material band edge. This means that the wavelength must be in the $0.8–0.9\,\mu m$ range (depending on Al concentration). These devices have shown the capability to operate at more than $10\,GHz$ [13] with energies of less than $5\,pJ$ [14]. Several logic gates have been demonstrated including XOR, NOR, and OR [15]. The capability of incorporating optical logic in a system can significantly improve the flexibility of the entire system.

Smearing of quantum wells of AlGaAs by diffusion of Zn or implantation of Si has been demonstrated and is under investigation in several laboratories. As the refractive index of the AlGaAs alloy is different than that of the original multi quantum well structure this effect provides an additional parameter that can be used for optical confinement, particularly in the lateral direction.

5.5.3 Optical Throughput Loss

Many forms of loss encountered in the waveguide can be related to a number of factors. These are material properties which include growth effects, the processing of the material for waveguide fabrication, and simply the design requirements. A reasonable fabrication but fairly stringent design tolerance for a GaAs/GaAlAs waveguide system might be $1\,dB\,cm^{-1}$ loss in straight sections for the waveguide and $0.3\,dB\,rad^{-1}$ loss in curved sections of the waveguide. This requires careful design and fabrication to minimize all contributors to loss, that is, interband absorption, free-carrier absorption, scattering, and radiation.

The major loss to be dealt with in an AlGaAs system is band-edge or interband absorption. In an ideal system, this loss can be overcome by simply using a waveguide of slightly less energy than the bandgap of the waveguide material less the energy of acceptors. In a real material with a source wavelength close to the band edge, a certain amount of loss is unavoidable due to deep level states. A large part of this loss can be attributed to band-edge tailing resulting from various defects and impurity states in the material. The quality of the material used for the waveguide can thus be directly correlated with the loss.

Using the early absorption data of Sturge [16] and Stoll et al. [17], one would project that interband absorption loss would be greater than

1 dB cm^{-1} for wavelengths shorter than 0.92 μm in waveguides with an aluminum concentration of 40%. More recent studies have measured losses of 3–5 dB cm^{-1} at wavelengths of 0.86 μm. The high quality material now available will further reduce losses to an acceptable level. It is actually inadvisable to reduce losses by going to aluminum concentration greater than 20% because the appearance of deep level defects becomes the limiting factor. At wavelength greater than 0.92 μm, waveguides with total propagation loss less than 1 dB cm^{-1} have been made. For example, Tracey et al. [18] reported MBE-grown waveguides with 1 dB cm^{-1} loss in either 1.06 or 1.15 μm waveguides.

Surface scattering from the walls of the waveguide can also be a significant loss mechanism if the roughness of the walls is not kept within certain limits. Tien [19] derived an expression for scattering loss due to surface roughness based on the Rayleigh criterion. The exponential loss coefficient α_s is given by

$$\alpha_3 = \frac{A^2}{2}\left(\frac{\cos^3\theta_m}{\sin\theta'_m}\right)\frac{1}{t_g + (1/p) + (1/q)} \tag{5.15}$$

where
   t_g is the thickness of the waveguiding layer
   θ_m is the angle between the ray of the waveguide light and the normal to the waveguide surface
   p and q are the extinction coefficients in the confining layers

The coefficient A is given by

$$A = \frac{4}{\lambda_2}(\sigma_{12}^2 + \sigma_{23}^2)^{1/2} \tag{5.16}$$

where
   λ_2 is the wavelength in the guiding layer
   σ_{12} and σ_{23} are the statistical variances of the surface roughness

Although Tien's expression was derived for the case of a three layer planar waveguide, it can be used to estimate the order of magnitude of surface scattering loss in a rectangular guide as well. Note that α_s is basically proportional to the square of the ratio of the roughness to the wavelength in the material (represented by A^2), weighted by secondary factors that take into account the shape of the optical mode. For the case of AlGaAs waveguides and wavelengths in the range of 0.82–1.3 μm, the wavelengths in the material are a few tenths of a micron; hence, if surface variations of the waveguide are limited to approximately 0.01 μm, α will be approximately 0.01 cm^{-1} and the loss due to interface scattering will be at most a few tenths of a dB cm^{-1}. Surface variations of less than 0.01 μm can be obtained in AlGaAs rectangular waveguides produced by either chemical etching, reactive ion etching, or sputtering etching. It should be possible to limit scattering loss even further in strip-loaded waveguides which have less sidewall scattering loss.

One technique that has been used to minimize the effect of surface roughness is to either deposit a thin passivation layer on the surface or to grow a thin GaAs or AlGaAs layer on top of the waveguide. These techniques may be necessary anyway to protect the surface from high field strengths and outside contamination [20]. The layers must be very thin (~0.1 μm) for devices that rely on the electro-optic effect because the overlap of the electric field will be reduced as the layer thickness increases.

Propagation loss due to free-carrier absorption is not a significant problem in AlGaAs waveguides because the carrier concentration can be adequately reduced in the waveguiding layer. The classical expression for the attenuation coefficient due to free-carrier concentration is [21]

$$\alpha_{fc} = \frac{Ne^3\lambda_0^2}{4\pi^2 n(m^*)^2 \mu \varepsilon_0 \sigma^3} \tag{5.17}$$

where
   λ_s is the vacuum wavelength of radiation being guided
   e is the electronic charge
   c is the speed of light in a vacuum
   n is the index of refraction
   ε_0 is the permittivity of free space
   m^* is the effective mass of the carriers
   u is their mobility
   N is the carrier concentration

For light in the wavelength range of 0.82–1.3 μm, traveling in an n-type doped AlGaAs waveguide, the absorption coefficient can be approximated by

$$\alpha_{fc} = 1 \times 10^{-18} N (\text{cm}^{-3}) \text{cm}^{-1} \tag{5.18}$$

Since a carrier concentration of $N = 10^{16}$ cm^{-3} can be routinely grown in AlGaAs layers with small aluminum concentration, one can expect α_{fc} to be on the order of 0.01 cm^{-1} or loss due to free-carrier absorption to be less than 0.1 dB cm^{-1}.

In curved sections of rectangular waveguides it is necessary to consider radiation loss in addition to the three loss mechanisms discussed previously. Radiation loss is theoretically unavoidable and increases greatly as the radius of curvature of the bend is decreased. The radiation loss depends exponentially on bend radius as given by the expression [22]

$$L = C_1 \exp\left(\frac{-2}{q} \frac{\beta_z - \beta_0}{\beta_0} R \right) dB\,cm^{-1} \qquad (5.19)$$

where
    C_1 is a constant that depends on the dimensions of the waveguide and on the shape of the optical mode
    R is the radius of curvature of the bend
    β_z is the propagation constant of the optical mode in the waveguide
    β_0 is the propagation constant of unguided light in the confining medium surrounding the waveguide
    q is the extinction coefficient of the optical mode in the confining medium

The key parameters in this expression are β_z, β_0, and R. In order to compensate for decreasing R it is necessary to increase B_z and/or to decrease B_0, that is, to increase the index of refraction difference between the waveguide and confining medium. Calculations done by Goell [23] for the case of rectangular dielectric waveguides of approximately 1.0 μm width indicate that an index change of only 1% is sufficient to reduce radiation loss to less than 0.1 dB cm⁻¹ for a bend radius of 1 mm. In summary, theoretical calculations and data published in the literature show that with AlGaAs waveguides, losses due to free-carrier absorption, surface scattering, and radiation loss can be kept within reasonable limits, while interband absorption loss represents the largest challenge. Other material systems may present loss problems of a greater magnitude than the AlGaAs system.

5.5.4 Material Growth: MOCVD versus MBE

Waveguide structures and devices are grown by a number of fabrication techniques, including the metal organic chemical vapor deposition (MOCVD) and molecular beam epitaxy (MBE) approaches. Both of these technologies have demonstrated excellent device results [24]. It is important to gather data and make comparisons between MOCVD and MBE for a particular device application in making a choice between the two growth techniques. In the following section some of the differences, advantages, and disadvantages of the two systems will be discussed.

    MOCVD presently has the lead in total output of the number of wafers that can be grown. The MOCVD technique is relatively well proven for the manufacture of a variety of AlGaAs system lasers and laser arrays [25]. MOCVD growth rates are four to five times faster than MBE growth rates; however, MBE manufacturers are developing sophisticated machines with multiple growth chambers capable of growing several wafers simultaneously. Since these growth chambers are completely isolated and contain their own individual material sources, one can grow different material systems simultaneously. MBE systems are much more complex than MOCVD and thus more susceptible to down time.

Material purity and defect density are two very important parameters that need to be considered when growing devices. In general, MBE demonstrated better control of background carrier concentration. MBE and MOCVD can achieve background carrier concentrations of less than 10^{15} cm^{-3} in AlGaAs. It is easier to achieve these low levels in GaAs epitaxial growth than it is in the ternary AlGaAs. One indication of the effect of background carrier concentration is to grow a multiple quantum well structure using both MOCVD and MBE and to subject the devices to high temperatures (~700°C) and measure the interdiffusion of the layers. This was performed by Hutcheson [26] and the results showed the MOCVD layers to have about 10 times more interdiffusion than the MBE structure. For a large number of devices (including lasers and passive waveguides) low background carrier concentration is not required. In fact, laser structures require carrier concentrations of 10^{18} cm^{-3}. High speed modulators and switches, however, do require background carrier concentrations of 10^{15} cm^{-3} or less. This is driven by the need to have high resistivity material when applying the voltages required to induce index changes via the electro-optic effect. When designing the material requirements for an optical circuit which includes modulators and switches one must specify the breakdown voltage to be much larger than required because all of the device processing after material growth tends to reduce the breakdown voltage by a factor of three to five.

Surface defect density has always been a problem for MBE growth while MOCVD has been able to overcome this particular problem. When comparing defect densities between MOCVD and MBE one must be cautious to compare only the relevant results. MOCVD can routinely produce defect densities of 10–50 cm^{-2} while, for optoelectronic devices, MBE has defect densities of 200–500 cm^{-2}. Although MBE material growers have achieved defect densities of less than 100 cm^{-2}, this was accomplished on MODFET and HEMT electronic structures. These structures are grown at low temperature (350°C) and are not suitable for optical devices. This is due to the traps that exist in the bandgap and cause excessive optical losses (>100 dB cm^{-1}). To get rid of these traps in the bandgap the layers must be grown at a higher temperature. In MBE, the higher the growth temperature, the larger the defect density. When integrating the large numbers of devices on a single substrate, yield is directly related to defect density. This may cause a severe limitation for the MBE technique.

MOCVD demonstrates better routine control of Al concentration and thickness than MBE. This means that if a design specified a certain Al concentration and thickness, MOCVD would usually come closer to meeting the specifications; however, MBE will in general have better uniformity over a full 3-in.-diameter substrate than MOCVD. Growth of AlGaAs on GaAs by the MBE process results in a better interface than the growth of the inverted structure, that is, GaAs on AlGaAs. This has been observed in a comparison of direct and inverted HEMT devices. Such a difference has not been reported for structures grown by the MOCVD technique. One final comparison between the two growth techniques involves monolithic integration of optical and electronic devices. This will probably require selective epitaxial growth, that is, growing well-controlled AlGaAs/GaAs layers on selected portions of the wafer. To date MOCVD has been more promising than MBE for selective epitaxy.

5.5.5 Microwave and Electronic Circuit Compatibility

The device structures and fabrication processes for AlGaAs waveguide devices are inherently compatible with those for microwave and electronic integrated circuits because the same materials and processing techniques are used. Both field effect transistors (FET) and bipolar transistors can be fabricated in AlGaAs structures on a GaAs substrate. In fact, high performance microwave transistors and high speed electronic switching transistors such as the modulation-doped FET (MODFET), the high electron mobility transistor (HEMT), and the heterojunction bipolar transistor (HBT), are all based on AlGaAs system heterojunctions.

Most electronic integrated circuits are currently fabricated in silicon because of the well-developed fabrication technology for that material. Research and development, however, has clearly demonstrated that GaAs integrated circuits are capable of much higher frequency operation because of the greater electron mobility and scattering limited velocity in GaAs relative to silicon. The fabrication processes used to make these GaAs integrated circuits are essentially the same as those used to make optical devices on GaAs substrates. Additional details regarding the monolithic integration of optical, microwave, and electronic devices on a single GaAs substrate are contained in the next section.

5.5.6 Integratability

When hybrid integration offers the advantage of expediency, fully monolithic integration of all microwave and optical elements on the same substrate to form an optomicrowave integrated circuit (OMMIC) is the ultimate preferred approach, because it offers the greatest reliability and the lowest cost once a high-volume production line is established. There are, of course, some problems of integratability that must be dealt with. Optical integrated circuits require low-loss transmission of optical signals, efficient generation and detection of optical power at the desired wavelength, and efficient modulation and switching of optical signals with a wide bandwidth. Microwave integrated circuits require low-loss transmission of microwave signals and minimum electromagnetic coupling (crosstalk) between various signal paths. It is important that the waveguide building blocks for an optical circuit be designed within a system framework that is compatible with the eventual monolithic integration of laser sources, photodetectors, microwave devices, and electronic elements.

The approach using epitaxial growth of AlGaAs optical and microwave devices on a semi-insulating Cr-doped GaAs substrate offers excellent compatibility with the requirements of both optical and microwave integrated circuits. The low-loss transmission of microwave signals via metallic stripline waveguides fabricated on Cr-doped GaAs substrates has been well documented and is widely used even in commercially available monolithic microwave integrated circuits (MMICs). The Cr-doped substrate has a resistivity of approximately 10^7 or 10^8 Ω cm^{-1} and behaves essentially as a high field strength, low-loss dielectric. Microwave losses and leakages are thus minimized. The use of such a substrate in an optical integrated circuit complicates the design somewhat. The semiconductor substrate cannot be used as a return ground path for the electrical current through lasers, photodiodes, and modulators. This problem can be solved rather easily, however, by incorporating a buried n^+ layer and by designing optical device structures that have all their electrical terminals on the top surface of the wafer. Efficient lasers, modulators/switches, and detectors of this type have been demonstrated [27]. Standard photolithographic metallization techniques can then be used to interconnect devices and power sources as in a conventional electronic integrated circuit. Metallization lines can be deposited directly onto the semi-insulating Cr-doped substrate without any intervening oxide layer because of the high substrate resistivity and dielectric strength. A deposited oxide or nitride layer can be used where intersections (crossovers) of metal lines occur, as long as the insulating layer is thick enough (typically 1.0 μm) to reduce capacitive coupling to an acceptable level.

Monolithic optical integrated circuits are commonly fabricated in AlGaAs epitaxial grown layers on GaAs substrates. The energy bandgap and index of refraction can be conveniently adjusted through control of compositional atomic fractions to produce efficient light emitters and detectors as well as low-loss waveguides, couplers, and switches. This technology is well developed, along with improved techniques of patterned epitaxial growth and etching to define the circuit elements in development.

Monolithic microwave integrated circuits are fabricated on semi-insulating GaAs substrate in epitaxially grown layers of GaAs to take advantage of the high electron mobility and scattering limited velocity in these materials. Microwave generators such as IMPATT and Gunn diodes can be fabricated as well as FETs for amplification and switching. Passive elements such as capacitors and inductors for filtering and impedance matching can be fabricated by metal film deposition directly onto the semi-insulating substrate. The circuit elements are defined by conventional photolithographic techniques. This MMIC technology is all basically compatible with that used for OIC fabrication. Care must be taken in the design of OMMICSs to avoid any undesired stray interactions between devices. For example, an IMPATT oscillator should not be located directly next to a laser diode because they are both high power heat generating devices that represent a "hot spot" on the wafer. Devices should also be separated far enough to prevent stray coupling through electro-magnetic fields. Spacings greater than 1 μm should be sufficient because both the optical and microwave devices are fabricated in thin layers (~10 μm), so fringing fields do not extend far beyond the device.

MMIC and OIC fabrication technologies are compatible to the extent that fabrication of an OMMIC on a GaAs substrate does not require any extensive modification of the processes already in place. Several examples of OMMICs have already been demonstrated [28]. Furthermore, the heterojunction waveguide structure is basically the same as that for a double heterojunction laser, that is, the device with a reverse bias operates as a modulator but, with a forward bias, it operates as a laser. Monolithically integrated FETs have been used to modulate AlGaAs laser diodes on the same substrate at GHz frequencies [29], and a monolithic integrated circuit consisting of microwave FETs, an AlGaAs laser, and a photodiode has been used to produce a feedback-stabilized laser light source [30].

5.6 Processing and Compatibility Constraints

5.6.1 Introduction

The fabrication process should be designed so that it is compatible with electronic and microwave device requirements as well as those for optical devices. The various steps of the fabrication process must be compatible in terms of temperatures required, chemical interactions, and structural strength considerations.

5.6.2 Substrate Specifications

The first step in the process design procedure should be the determination of specific requirements for the substrate material. It is clear that Cr-doped semi-insulating gallium arsenide will provide electrical isolation between devices and will make possible microwave stripline interconnections for compatibility with the eventual incorporation of monolithic microwave devices. Additional constraints on the substrate material will have to be determined. For example, ideally, the substrate material should contain just enough chromium atoms to compensate the residual background dopant concentrations (usually about 10^{15} cm^{-3} Si atoms in the highest purity GaAs, which acts as donors). Excess Cr atoms can diffuse into the epitaxial layer during the growth producing a nonuniformly doped compensated layer with lower carrier mobility and increased optical scattering. Also, the surface polishing normally provided by the suppliers of GaAs wafers or electronic and microwave integrated circuits may not be smooth enough for the fabrication of optical components with low scattering loss. Additionally chemo-mechanical polishing to remove surface damage generally found in "polished" wafers may be necessary.

5.6.3 Epitaxial Growth

One must determine whether it is better to use oxide or nitride masking techniques to define the lateral dimensions of devices during epitaxial growth, or to grow a layer over the entire wafer and then mask and etch away unwanted portions of the layer. The desired percentages of constituent elements and dopants in each layer must be determined to control index of refraction, optical absorption, and scattering for optical devices, along with regard for carrier concentration, mobility, and lifetime for electronic and microwave devices. Fortunately, many of the requirements are the same for all of these devices. For example, low defect densities in the epi-layer which are necessary for minimal optical scattering, are also required for high carrier mobility and scattering-limited velocity in microwave and electronic devices. Accurate control of dopant densities, which is needed to produce low background carrier concentration in order to minimize free-carrier absorption, also results in reduced microwave loss and permits the fabrications of microwave IMPATT and Gunn diodes.

5.6.4 Metallization

The metallization material and deposition technique chosen should be compatible with microwave, electronic, and optical devices. Most of the metallization requirements are the same for all these devices, that is, high electrical conductivity, good adherence to GaAs and related III–V ternary and quaternary materials, no adverse chemical reactions, accurate pattern definition, and ease of deposition. In some optical devices it may be desirable to utilize a semitransparent metal film. Thus the chosen metal should be one that can be deposited in very thin layers (~1000 Å) for transparency as well as in layers of several microns thickness to produce high conductance. The most likely metallization system to satisfy these requirements on GaAs is the widely-used chrome-gold technique (a very thin layer of Cr to provide reliable adherence, capped with a thicker layer of Au for high conductance). Other potentially useful metallizations such as aluminum may also be considered.

5.6.5 Thin-Film Insulators

Thin-film insulators may be required at various points to provide electrical insulation for metallic crossovers. The material chosen should have a relatively large dielectric constant and be able to be deposited in thin (1000 Å) pin-hole free layers to facilitate its use as an insulator in capacitors in electronic or microwave circuits. The insulators should have an index of refraction less than that of GaAs to prevent stray optical coupling (all of the commonly-used oxides and nitrides satisfy this requirement). In addition, the insulating material should be available in high purity form, and the chosen deposition technique should not introduce significant amounts of any atoms that can act as dopants in GaAs. For example ion-beam sputter-deposited SiO_2 layers deposited by thermally excited vacuum vapor deposition would not be a problem. More than one insulating material might be used on the same substrate wafer to take advantage of differences in index of refraction, microwave dielectric constant, or differing processing characteristics such as chemical etch rate. Nonetheless, the thermal expansion coefficients of all layers should be adequately matched to provide good adherence.

5.6.6 Photolithography

It is possible to fabricate common optical, electronic and microwave devices by standard photolithographic techniques. The resists and exposure methods used for microwave integrated circuit fabrication are essentially the same as those used to produce optical integrated circuits. The one element which is an exception is the optical grating used for coupling or feedback. Spacing of the grating bars can be as small as 0.1 μm, which is within the resolution of commonly used photoresists, but is beyond the capabilities of conventional mask alignment machines. If gratings are used in any of the optical devices it will be necessary to expose the photoresistors for the grating by interference (sometimes called "holographic") techniques [31]. Most other optical, microwave, and electronic devices have minimum dimensions greater than 1.0 μm, which can be produced by commercially available mask alignment machines.

5.7 Waveguide Building Block Processing Considerations

5.7.1 Introduction

As discussed in a previous section, the buried strip-loaded waveguide has some inherent advantages over the simple raised rib design. First, the vertical confinement is such that the mode energy is generally contained in the waveguide layer and thus only sees the minimal roughness of the interfacial layer as opposed to the etched sidewalls. Second, a structural design analysis shows that lateral confinement is achieved with a very reasonable strip width. Third, from an electrical point of view, the p–n junction is buried and therefore will exhibit forward electrical breakdown characteristics when making electro-optic modulators or lasers, because the electric field is a maximum in the optical waveguiding layer. Optimum coupling of the electrical and optical fields is obtained through the electro-optic coefficient. Fourth, the heterostructure design naturally leads to the integration of lasers, waveguides, and modulators without undue etching of material.

5.7.2 Material Systems: Control of Loss, Refractive Index, and Electro-Optic Effect

By choosing the AlGaAs system for waveguide fabrication, one is essentially obtaining a wide range of freedom in the choice of band gap for the material. By varying the percentage of the Al the band gap of the material may be changed significantly. The band gap is directly related to the emission and absorption wavelengths of the material. Figure 5.16 shows the band gap and related wavelength and photon energy of the AlGaAs system for varying percentages of Al from 0% (GaAs) to 40%. Figure 5.17 shows the index of refractions of AlGaAs versus Al concentration for several wavelengths. It is seen that by a proper choice of material composition in the waveguide core and cladding regions a suitable guiding structure may be designed.

The strip ridge should be aligned with either the (011) or the (110) crystallographic direction for a modulating electric field perpendicular to the substrate (100) direction. The change in refractive index due to the electro-optic effect is

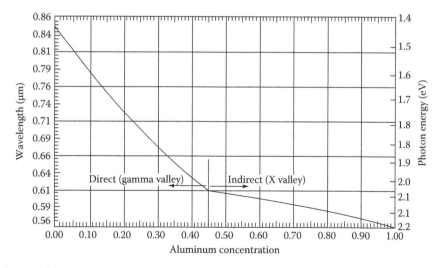

FIGURE 5.16
Minimum energy gap versus aluminum concentration.

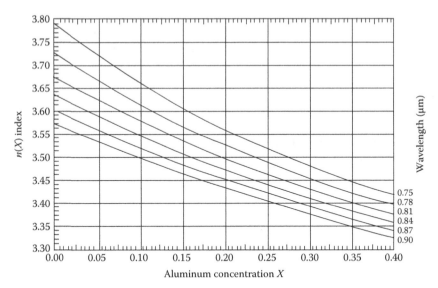

FIGURE 5.17
$Al_xGa_{1-x}As$ $n(X)$ versus X and vacuum wavelength.

$$n = \frac{-\alpha n_0^3 r_{41} E}{2} \quad \text{for (011) propagation} \tag{5.20}$$

where

r_{41} is the electro-optic coefficient (for GaAs $r_{41} = 1.1 \times 10^{-12}$ m V^{-1})
α is an overlap figure with a value between 0 and 1 depending on the overlap of the electric and optical fields

One way to get a larger field overlap is to grow highly conducting n^+ layer on top of the SI GaAs substrate before growing the buffer layer. This causes the electric field to be perfectly vertical. It is apparent from the equations that because of the small size of r_{41} a device structure which can sustain a high electric field is a necessity if unduly long devices are to be avoided.

5.8 Coupling Considerations

5.8.1 Fiber to Waveguide Coupling

The problem of coupling an optical fiber to a waveguide (or a waveguide to a fiber) is critically important to the overall performance of integrated/fiber-optic systems. The most straightforward way of coupling a fiber and a

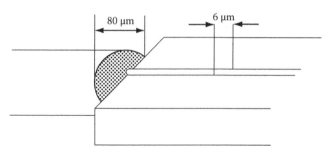

FIGURE 5.18
Butt coupling of fiber to integrated optic waveguide.

waveguide is the end-fire or butt coupling technique in which the two are aligned end to end and butted into contact (see Figure 5.18). The key factors affecting butt coupling between a fiber and a waveguide are area mismatch between the waveguide cross-sectional area and that of the fiber core, misalignment of the waveguide and fiber axes, and numerical aperture loss (caused by that part of the waveguide light output profile that lies outside of the fiber's acceptance core or vice versa). In the case of butt coupling of the semiconductor waveguide and a glass laser, reflection at the glass-air-semiconductor interface are also important.

In considering coupling of light from the fiber into the waveguide, the problems of area mismatch and numerical aperture loss are limiting factors because the core diameter of a single mode fiber is approximately 4–10 µm and the rectangular waveguide dimensions are 1–5 µm. Exact calculations of the coupling efficiency from fiber to waveguide cannot be done until a particular waveguide and optical fiber have been selected and their characteristics known; however, a typical experimentally observed coupling efficiency for this situation would be about 10%. Robertson et al. [32] have performed a theoretical analysis of butt coupling between a semiconductor rib waveguide and a single mode fiber, and have designed an optimized waveguide structure with a large (single) mode size for which they calculate a coupling efficiency of 86%. However, no experimental data is available.

One approach to the problem of coupling fibers to multiple, closely spaced waveguides is to fabricate a "v-groove chuck" with etched grooves to position and hold the fibers in place, as shown in Figure 5.19. Due to the precision of photolithographic techniques used to etch the grooves and the dimensional precision of standard optical fibers, this method provides a highly accurate means of aligning the fibers and waveguides.

In principle, grating couplers can be used to avoid the problem of area mismatch and numerical aperture loss in coupling between a fiber and a waveguide by providing a coherent distributed coupling over a mutually coupled length of fiber and waveguide. For example, Hammer et al. [33] demonstrated that a grating can be used to couple between a low-index fiber

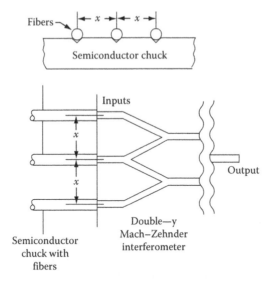

FIGURE 5.19
Fiber input coupling using semiconductor chuck.

and a high-index waveguide. By embossing an unclad multimode plastic (PMMA) fiber ($n = 1.5$) over a grating etched into a diffused $LiNb_xTa_{(1-x)^3}$ waveguide ($n = 2.195$), they observed a 6% coupling efficiency. Their calculations indicated, however, that 86% efficiency should be obtainable with an optimally blazed grating. Bulmer and Wilson [34] also used a grating to couple between a single mode fiber and a sputtered glass (Corning 7059) waveguide. In that case the fiber was heated and drawn to reduce the cladding thickness to the point where coherent coupling through the overlapping evanescent mode tails could occur. Again the observed coupling efficiency was low (0.4%), but the authors calculate they could have obtained 30% efficiency if they had used a fiber with a core diameter equal to the width of the waveguide.

5.8.2 Waveguide to Fiber Coupling

Although the same limiting factors are relevant, butt coupling of light from the waveguide to the fiber is an easier task than the converse because the area of the fiber core is larger than the waveguide cross sectional area. In that case a lens can be used to couple more of the light from the waveguide into the fiber. By using a proper intervening lens the effective solid angle of collection can be increased by a factor equal to the ratio of the area of the fiber core to that of the waveguide. By using a cylindrical lens to better match the rectangular waveguide mode to the circular fiber a coupling efficiency of at least 50% should be obtainable [35].

To obtain efficient coupling between optical fibers and waveguides very accurate alignment of the fiber and waveguide is required. Fiber positioning systems are available commercially that provide repeatable accuracy of .1 µm. Automated, production-type, alignment machines that provide 0.1 µm resolution are also available.

5.8.3 Laser Diode to Waveguide Coupling

Hybrid coupling of a discrete laser diode to a semiconductor waveguide can be efficiently done by means of the butt coupling technique. Theoretical studies of laser-waveguide butt coupling [36] indicate that coupling efficiency to a single mode waveguide can exceed 90% if the refractive indices and effective cross-sectional areas of the laser light emitting layer and the waveguide are well matched. Alignment tolerance of less than 0.1 µm is required; however, commercially available piezoelectrically driven X–Y–Z micropositioners are available at reasonable cost that easily meet this requirement. Reliable hybrid butt coupling of an AlGaAs laser diode to lithium niobate waveguide has been demonstrated in the Integrated Optic RF Spectrum Analyzer program [37]. Butt coupling between single mode semiconductor lasers and strip waveguides with a theoretical coupling efficiency of greater than 95% has been predicted by Emkey [38].

Because of the convenience and demonstrated effectiveness of end-butt coupling between the laser and waveguide, it should not be necessary to consider more complicated methods, such as those involving mirrors, prisms, and lenses, for most applications. In addition to being an effective method for coupling laser sources for testing and evaluating waveguide devices, end-butt coupling represents a close approximation to the ultimate coupling of a laser diode monolithically fabricated on the same substrate as the semiconductor waveguide. Thus, evaluation of the characteristics of the hybrid-coupled laser/waveguide combination should provide a good indication of what to expect in the monolithic case.

5.8.4 Waveguide to Detector Coupling

A photodiode detector can be butt coupled to the semiconductor waveguide in the same way described for the laser diode. Ideally one would like an edge-oriented photodiode such as would be used in a monolithically fabricated OIC. Thus, for convenience in testing waveguide devices it is often best to use commercially available, surface-oriented germanium photodiodes. These photodiodes generally have surface areas that are relatively large compared to the cross sectional area of the waveguide; hence, direct end butt coupling results in low coupling efficiency. In this case an intervening lens can be used to compensate for the area mismatch to produce efficient coupling. The 0.7 eV bandgap of Ge results in an absorption edge at 1.55 µm. Thus, a Ge photodiode can be used throughout the wavelength range from 0.8 to 1.3 µm.

5.9 Lithium Niobate Technology

Titanium diffusion into lithium niobate ($LiNbO_3$) is a process used for fabricating waveguides and other devices of an optical circuit. Although $LiNbO_3$ is not an active semiconductor material such as GaAs and hence does not afford the monolithic integration of sources and detectors, much of the work in optical circuit engineering to date has been performed in this material due to the high electro-optic and acousto-optic figures of merit, ease of processing, high optical transparency (<0.1 dB cm^{-1} total loss from 0.4 to >2 µm), chemical stability, and lack of the requirements for high-level integration and monolithic compatibility. As the ancillary technologies continue to mature, GaAs and semiconductor based optical circuits will continue to supplant the $LiNbO_3$ technology; nevertheless, many of the emerging commercial communications, to be discussed in later sections, are based in $LiNbO_3$.

Selection of the operating wavelength for a $LiNbO_3$ system depends upon many of the same criteria as other material systems. 1.3 and 1.5 µm correspond to the lowest absorption loss wavelengths for silica-based fibers (0.5 and ~0.2 dB km^{-1}) [39]. The variation of refractive index dispersion which limits the potential modulation rate is also minimized at these wavelengths, so that many of the devices currently available are designed for operation in these regions, along with the AlGaAs source wavelength. At this time the cost and performance of emitters shifts the balance in favor of 1.3 µm except in cases of very long fiber lengths where the lower attenuation at 1.55 µm outweighs these factors. Fluoride fibers currently exist which are capable of less than 1 dB km^{-1} and theoretically as low as 10^{-2} dB km^{-1} at a wavelength of 2.55 µm. Research on InGaAs detectors indicates they make good detectors at that wavelength [40]. Fabrication of these devices is relatively difficult and the cost of fluoride based fiber is higher than silica fibers so the implementation of such a system appears to be downstream.

5.9.1 Electro-Optic and Photorefractive Effects

The electro-optic coefficient translates into the control voltage required for an optical component to function. Typically $LiNbO_3$ devices require very low control voltages with key relations being the following:

Change in refractive index [41]

$$\Delta\left(\frac{1}{n^2}\right) = rE + RE^2 \tag{5.21}$$

where
  E is the applied electric field (voltage/electrode spacing)
  r is the Pockels effect coefficient
  R is the Kerr effect coefficient

Optical phase retardation [42]

$$\Gamma = \frac{2\pi l}{\lambda_0}(n_1(E) - n_2(E)) \tag{5.22}$$

where
$\Gamma = Ed/V$, l = crystal thickness in the direction of the applied field
$n_1(E)$, $n_2(E)$ are the field dependent indices of refraction

The voltage length for a phase shift $V\pi L$ is

$$V\pi L = \frac{d}{n^3 r} \tag{5.23}$$

This is an important parameter as it incorporates the degree of overlap between the optic and electrical fields and must be minimized for low voltage operation.

Another property of LiNbO$_3$ which may be exploited for some devices is the photorefractive effect (a permanent change in the refractive index profile caused by the transmitted optical flux). Optical power levels must be limited to prevent refractive index changes. The effect can be alleviated to a certain degree via the use of longer wavelengths, or it can be exploited in device applications. Integrated optic lenses and reflectors can be fabricated through control of the refractive index. The effect is controllable in LiNbO$_3$ crystals as a function of the impurity levels present [43]. The field spatial change will be of the form

$$E = \frac{1}{\varepsilon}\left[k\alpha l(z) + eD\frac{dn}{dz} \right]dt \tag{5.24}$$

where
ε is the permittivity
k is the photovoltaic coefficient
α is the absorption coefficient
$l(z)$ is the intensity of incident light
D is the diffusion coefficient
e is the charge of an electron
dn/dz is the concentration gradient of free carriers

The first term relates to the ionization of impurities present in the LiNbO$_3$, while the second term describes the current induced by the diffusion of free carriers. Whether the photorefractive effect is a detriment depends on the device application and design.

5.9.2 Photolithography and Waveguide Fabrication

The first LiNbO$_3$ waveguides reported were produced via the outdiffusion of lithium oxide from the crystal lattice from photolithographically defined regions. Although this method was effective at changing the refractive index of selected regions to define the waveguide it tended to produce guides with high losses, and it was only possible to produce multimode waveguides [44]. The high losses have been attributed to lattice imperfections resulting from the outdiffusion [45]. The first devices formed via titanium indiffusion were reported in 1974 [46] and represented a substantial improvement over the existing technology. Devices based on the Ti diffusion process also provide greater possible changes in the index of refraction and less in-plane scattering [47] compared to the lithium oxide out diffused devices.

The fabrication of waveguides in LiNbO$_3$ materials via Ti diffusion is a relatively straightforward process and employs the sequence shown in Figure 5.20 [48]. After surface polishing, areas are defined on the LiNbO$_3$ surface where waveguides are desired, typically by a photolithographic process. Next Ti is deposited over the entire substrate; then, the resist is removed, lifting off the deposited Ti from all but the defined waveguide areas. The Ti is then diffused into the crystal at high temperatures under controlled conditions. Precise descriptions of these individual steps, their significance with regard to device performance, and ongoing research in the related photolithography and diffusion physics will be discussed in the following sections.

The first step in the processing of LiNbO$_3$ for waveguides is surface preparation. The crystal surface must be very flat and low in camber in order to enable the subsequent fabrication of devices with good yields. One of the advantages of LiNbO$_3$ over the other electro-optic materials arises from the fact that it is a strong nonhydroscopic crystal which is easily polished [49]. Another advantage is its availability. Currently high quality substrates of LiNbO$_3$ up to 8 cm long are commercially available [50] so there are cost advantages tied to the use of this material. There are several other ferroelectric materials which offer similar or even slightly better properties [51] but when cost and availability are taken into account LiNbO$_3$ is the preferred choice [52].

After polishing, the LiNbO$_3$ surface is ready for liquid photoresist application. These photoresists are polymeric solutions which are sensitive to certain types of (typically UV) radiation. The polished surface can be coated with resist via spray or dip application, but spin coating is the most common method. In spin coating the substrate is turned very rapidly, up to 5000 rpm, on a turntable as a precisely measured amount of photoresist is poured onto the center of the spinning substrate. This results in a thin, very uniform thickness coating across the surface of the substrate. These resist solutions are only 20%–33% solid so there is a drying step required to drive off all the carrier solvents.

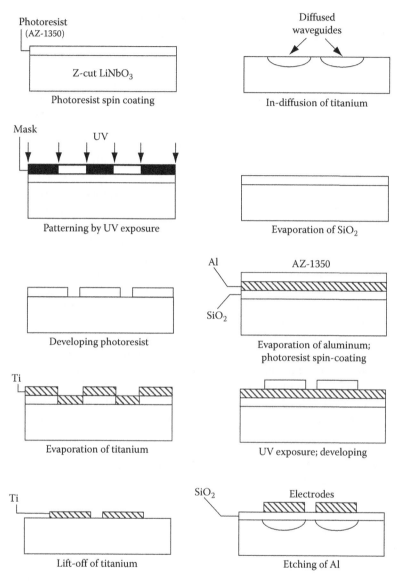

FIGURE 5.20
Fabrication for channel waveguides using Ti:Indiffused LiNbO$_3$.

After drying the resist is ready for exposure. There are two basic types of photoresists: negative acting and positive acting. Both types function via altering of the polymer solubility due to exposure. With exposure in selected areas some regions will be soluble in the developing solutions while others remain insoluble and form the resist surface. In negative acting resists exposure causes polymerization of the material making the exposed regions insoluble. In positive acting resists exposure causes depolymerization, making the exposed areas soluble and thus removable in suitable solvents.

Both resists have some advantages. Negative resists tend to be denser, stronger coatings which stand up better to harsh processing steps. This is because the exposure causes the formation of an interpenetrating polymer network with a much greater degree of crosslinking than in the positive resists. Positive resists generally produce slightly better resolution. This is because they do not suffer the slight shrinkage resulting from crosslinking, nor are they prone to swelling in the developing solutions as with negative resists.

For the standard Ti diffused LiNbO$_3$ waveguides positive resists tend to dominate; however, for some of the newer processes which require a photoresist with greater chemical resistance the negative resists are preferred. Line width control is still a major concern but advances in negative resist chemistry put them almost on a par with positive. Limitations for negative resists under optimal conditions are currently 2.0 µm while for positive resists they are 0.8 µm. Resolution is critical in Ti:LiNbO$_3$ devices in that control and definition of channel width are key to proper waveguide functionality. This is especially true for devices such as Bragg cells where diffraction efficiency is a direct function of waveguide geometry control.

Traditionally both types of resists have required exposure through a suitable patterned artwork or phototool. These phototools contain the desired circuitry or waveguide pattern dependent on the resist type being used. The phototools are typically made of chrome on glass substrates, are very stable with respect to temperature and humidity, and are quite capable of providing the degree of resolution required. They are, however, fragile and expensive as well as time consuming to produce. Often the production of hard tooling can be the most costly and time consuming part of the manufacturing process. An emerging technology that may eliminate the need for phototools is that of direct write lithography. Due to their complexity and density all of the required patterns for producing waveguides are currently generated using computer automated design. This design is then translated from the computer to produce the phototool which is then used to image the pattern onto the substrate. All of these steps represent costs and potentials for error and losses in resolution due to transferring the image from one source to another. In direct write lithography the computer containing the artwork pattern controls a suitable exposure source which is used to directly write the desired pattern onto the photoresist coated substrates without the use of a phototool. In theory these systems could reduce the costs and turnaround times dramatically of prototype, custom, and short run devices by eliminating the need to fabricate hard tools.

There are two methods which may be used to write images: raster scan and vector scan. In raster scan the exposure beam is scanned back and forth across the entire substrate. The pattern is written as a series of pixels by turning the beam on and off as required to define the pattern. In vector scanning the beam only traverses those areas to be exposed. In practice vector scanning has proven to be somewhat faster at scanning a substrate; however, it does require the use of a shaped beam which limits its power compared to raster scanned beams [53].

There are several different types of direct write exposure equipment available, including excimer laser, ion beam, electron beam, and Argon laser system. At this time there is not a preferred system, as all of the systems are capable of producing ultrafine lines and good aspect ratio (height to width ratios), but all are quite slow. Most photoresists currently available are sensitive to radiation in the range of either 365 or 248 nm. This is because most of the optical lithography exposure equipment available exhibits its strongest output peaks at one of these values. Currently available resists require relatively high levels of exposure so that improved resists are required that better respond to the outputs of the direct write systems.

Laser systems using argon laser sensitive resists are in development for certain applications. The new resist films are expected to require only ~1–5 mJ of energy for exposure, which is about 1/10–1/100 of the current requirement. Another area of research involves the development of a new class of resists based on laser volatilized polymers. These are materials which can be vaporized to CO, CO_2, and H_2O via exposure to a laser source. The resist coating would be processed as described up to the exposure step, but the laser would selectively destroy the resist as the pattern required. The chief advantage to such a system is that it would eliminate the need for a developing step.

After exposure, development, and a final bake (to increase crosslink density) the imaged $LiNbO_3$ substrate is ready for further processing. The next step is the coating of the piece with titanium. This may be accomplished via resistance heated evaporators, sputtering, or electron beam process (see Figure 5.21) [54]. This step produces a uniform metal layer deposited over the entire

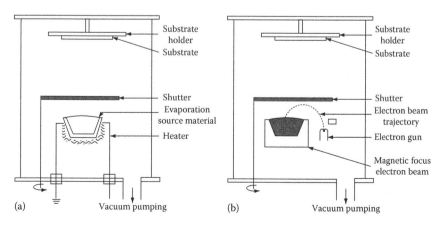

FIGURE 5.21
(a) Resistance evaporation and (b) electron beam evaporation.

piece. The thickness of the Ti coating is determined by the deposition rate and contact time. The coating thickness is critical in that it is a key parameter in determining the diffusion depth and concentration. Typically the metal will be deposited to a height of 0.3–1.0 μm depending on the wavelength of operation and the degree of confinement of the waveguide. Ti metal will be in direct contact with the LiNbO$_3$ only in those channels opened up by the photolithographic process. The next step is the lift off or stripping of the resist. The Ti coated piece is immersed in a suitable solvent that attacks the polymerized resist causing it to disintegrate and the Ti adhering to it to lift or wash off. The resist's aspect ratios become important at this point in that the resist must be substantially thicker than the Ti coating. If not, the metal coating adhered to the LiNbO$_3$ can bridge over and become adhered to metal covering the resist, which results in a phenomena known as mushrooming (see Figure 5.22). This can result in poor line definition and ultimately in poor device performance.

Titanium indiffusion is the next key area in the fabrication of a LiNbO$_3$ waveguide device. Extensive research continues in this area to characterize more fully the exact nature of Ti diffusion into the material and to more predictably control the diffusion. The current preferred process is to indiffuse the Ti at a temperature of 980°C–1050°C for 4–12 h. The diffusion must be performed under tightly controlled atmospheric conditions to prevent the outdiffusing of LiO$_3$ which can accompany metal indiffusion. The first LiNbO$_3$ waveguide devices were fabricated by taking advantage of the outdiffusing of LiO$_3$ at high temperature [55]. With the current technology this outdiffusing is considered to be a major problem as it can result in the formation of surface waveguides outside the channel regions [56].

Molecular diffusion is a means for mass transfer which results from the thermal motion of molecules and is limited by collisions between molecules. The first step in understanding the diffusion process is to develop an understanding of Fick's Law

$$dn = -Dq\left(\frac{d\sigma}{dx}\right)dt \tag{5.25}$$

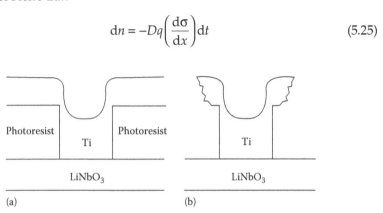

(a)   (b)

FIGURE 5.22
(a) Ti coated over photoresist with and insufficient aspect ratio and (b) "Mushrooming" noted after resist stripping.

for one dimension, and

$$\frac{\partial \sigma}{\partial t} = \frac{\partial}{\partial x}\left(D^{(xyz)}\frac{\partial \sigma}{\partial x}\right) + \frac{\partial}{\partial y}\left(D^{(xyz)}\frac{\partial \sigma}{\partial y}\right) + \frac{\partial}{\partial z}\left(D^{(xyz)}\frac{\partial \sigma}{\partial z}\right) \qquad (5.26)$$

for three dimensions, where dn = amount of a substance diffusing, q = cross section (along x-axis), $-dc/dx$ = concentration gradient, dt = time, and D = diffusion constant. This states that an amount of some substance (dn) will diffuse through a cross section (q) along the x-axis in a time (dt) so long as a concentration gradient exists, and that the amount is governed by a diffusion constant. As the concentration gradient goes to zero (i.e., complete uniformity) the diffusion falls off to zero. Diffusivities in solids cover a wide range of values and are typically quite low for dense crystalline materials such as $LiNbO_3$. A material diffusion constant (D) is not a universal constant but is rather a constant for specific situations and conditions. Diffusion coefficients are unique to individual materials and are a strong function (usually linear for solids) of temperature. For crystalline materials such as $LiNbO_3$ multiple diffusion coefficients exist depending on the plane of interest [57]. As molecular diffusion is limited by collisions between molecules these planar variations are expected.

Fick's Law is merely a general statement of diffusion theory and, although it cannot be violated, many specific forms of the law exist and apply to different situations. For diffusion of Ti into $LiNbO_3$ the diffusion depth, as expected from Fick's Law, is a function of the material diffusion constant, the diffusion time, and the diffusion temperature. The diffusion profile has been found to be roughly exponential, and more specifically the Ti concentration has been found to follow a Gaussian distribution [58]:

$$c(x_3, t) = \frac{\tau}{(4\pi D_y t)^{0.5} \exp\left[-x_3^2(4d_y t)\right]} \qquad (5.27)$$

where
    $c(x_3, t)$ is the Ti concentration in the $LiNbO_3$ crystal
    t is the diffusion time
    τ is the thickness of Ti coating prior to diffusion
    D_y is the anisotropic diffusion coefficient (temperature dependent)

The general form of the temperature dependent diffusion coefficient is

$$D = D_0' e^{[-E/kT]} \qquad (5.28)$$

where
    T is the crystal temperature
    E is the activation energy

For Ti diffusion into $LiNbO_3$ the temperature dependent D has been found exponentially to be [59]:

$$D = D_{y0}e^{[-T_0/T]} \qquad (5.29)$$

where
   T is the absolute temperature
   $T_0 = 2.9 \times 10^4$
   $D_{y0} = 7 \times 10^{-3}$ cm^2/s

The form appears logical since E can be expected to be quite low at the surface, yet increasing with diffusion depth. For this type of diffusion, depth dependent diffusion coefficients are being pursued via mathematical modeling [60]. These are particularly valuable in explaining the lateral surface diffusion which is often observed in the fabrication of these devices [61]. One particular phenomena which greatly benefits these devices results from the lower observed lateral diffusion rates. It has been noted that the diffusion of Ti down (vertical) into the $LiNbO_3$ crystal is about 1.5 times the horizontal diffusion rate (dependent upon the orientation of the crystal) [62]. This enhances the quality of the channels being formed and ultimately the devices being fabricated.

As seen from the equation the Ti diffused into the crystals is a linear function of Ti thickness; thus the Ti thickness can be used to determine the refractive index change. This assumes that all of the Ti is indiffused during the diffusion time (t). A proposed equation for this relation is [63]

$$\Delta n = \left(\frac{2}{\sqrt{p}} \alpha \frac{dn}{dc} \right) \frac{\tau}{D} \qquad (5.30)$$

where
   dn/dc is the change in refractive index with concentration
   α is a proportionality constant

This change has been found to be $dn_e/dc = 0.47$ and $dn_0/dc = 0.625$, where n_e and n_0 are extraordinary and ordinary refractive indices. For a specific wavelength the diffusion depth (D^*) is of the form

$$D^* = 2\left[D_{y0}te^{(-T_0/T)^{0.5}} \right]$$

$$D^* = 2(Dt)^{0.5} \qquad (5.31)$$

This shows that all the important parameters for defining the waveguide may be controlled within fabrication. Waveguide width is defined via photolithographic techniques, depth is controlled via diffusion time, and change in index of refraction of the waveguide versus the surrounding crystal is a linear function of the deposited Ti thickness.

There is an additional process step involved in the actual mechanism of diffusion. It has been discovered that Ti does not directly diffuse into the $LiNbO_3$ crystal but rather goes through an intermediate oxide phase before entering [64]. Important intermediate products are TiO_2, $LiNb_3O_8$ and $(Ti_{0.65}Nb_{0.35})O_2$ [65]. It has been noted experimentally that if the deposited Ti layer is first oxidized at temperatures of 300°C–500°C for approximately 4 h the resultant waveguides display better performance characteristics with regard to energy loss due to in-plane scattering [66]. It has been noted that this oxidation step should be done in an O_2 or dry air atmosphere [67] to prevent the excess stripping of O_2 from the $LiNbO_3$ crystal, and to prevent the formation of faults within the crystal [68]. It has been found that following complete TiO_2 formation a $LiNb_3O_8$ phase forms [69]. These two may then combine to form $(Ti_{0.65}N_{0.35})O_2$. The latter ternary compound begins to form at temperatures above 700°C and its formation rate increases up to 950°C. This compound has been identified as the true source of Ti for diffusion into the $LiNbO_3$ crystal, the ultimate compound being $(LiNb_3O_8)_{0.75}(TiO)_{0.25}$ wherein counter current diffusion of Nb^{+5} and Li^{+1} and indiffusion of Ti^{+4} occur.

The $LiNb_3O_8$ appears to be the resultant of Li or LiO_2 outdiffusion, and will form at temperatures of 650°C $< T <$ 950°C regardless of whether or not Ti is present. As stated previously outdiffusion of Li or LiO_2 causes changes in the index of refraction. The uncontrolled outdiffusion of these compounds in Ti indiffused waveguides is a major concern. This is because if left unchecked surface waveguides (the equivalent of electrical shorts) could form and limit the usefulness of the device. A number of studies have been undertaken [70] on the nature of structural faults which occur during the formation of Ti:$LiNbO_3$ waveguides. What is typically noted is that Li does outdiffuse and Nb molecules take their place in the waveguide portion of the crystal. Vacancies exist where the Nb had been. The faults are present in undoped (no Ti) Li deficient waveguides as point defects. In Ti indiffused systems they tend to form structural faults on the order of tens of microns [71]. Faults of this size are possible scattering sights for photons and thus their formation must be controlled. The growth kinetics of $LiNb_3O_8$ may be controlled via annealing of the waveguide in an atmosphere rich in Li in the presence of O_2 or in an atmosphere of steam [72]. Typical device parameters for the formation of a Ti:$LiNbO_3$ waveguide device would require channel widths of 8–10 μm, a coated metal (Ti) thickness of 0.3–1.0 μm, diffusion temperature of 980°C–1050°C, and a diffusion time of 4–12 h. This results in diffusion depths on the order of 5–10 μm. This process is used to form waveguides for operation in the 1.3–1.55 μm wavelength range where common silica-based fiber exhibits its best properties. Propagation losses as low as 0.2 dB cm^{-1} have been noted in waveguides produced via these techniques [73]. After diffusion the crystal end faces are polished to allow for coupling to a fiber. Waveguide devices for operation at a wavelength of 2.6 μm for use with the ultra low-loss fluoride based glass fiber have also been developed [74]. For these devices the optimal channel width is 13–15 μm, the required thickness is 0.8 μm, and the diffusion conditions are 1050°C for 12 h.

5.9.3 Implantation and Proton Exchange Techniques

A great deal of research continues with the goal of supplementing or replacing Ti diffusion. One technique is to ion implant Ti into the LiNbO₃ crystal. The waveguides regions are defined lithographically and implanted. This causes an amorphous layer to form in the waveguide regions. Solid phase epitaxy is then used starting from the undamaged underlying LiNbO₃ substrate in order to restore crystallinity and produce a functional waveguide. This process produces good waveguides with higher Ti concentrations and, therefore, greater changes in index of refraction than are possible via thermal diffusion [75]. It also promises sharper geometries and tighter Ti concentration gradients than are otherwise possible.

Another means for altering the index of refraction of LiNbO₃ is proton exchange. In this process the crystal (with defined waveguide regions) is submersed in a solution of benzoic acid at a temperature of 210°C–245°C for 1–4 h. The benzoic acid is the proton (H⁺) source and the diffusion of protons into the substrate causes the index of refraction to rise in those areas [76]. The process has the advantage of being a simple low temperature process, and is capable of achieving greater changes in the index of refraction than with Ti indiffusion. The drawback is that devices fabricated in this manner typically cannot achieve the same performance levels (losses) as LiNbO₃ devices. Annealing after the proton exchange improves the optical quality but also causes a drop in the achieved index of refraction.

Another exciting fabrication technique is known as Titanium Indiffused Proton Exchange (TIPE). This technique has the capability to produce devices not possible by either TI of PE processes alone. In TIPE a standard Ti indiffusion process is first applied to the LiNbO₃ substrate to define some regions with higher indices of refraction. The piece is then masked with a suitable image and immersed in a benzoic acid solution. This results in the areas of the crystal exposed to both Ti indiffusion and PE having relative indices of refraction considerably greater than achieved by either process alone. The most important and immediate application for this technology is in the fabrication of planar waveguide microlenses [77] (Figure 5.23a and b) and microlens arrays. Desirable properties of TIPE lenses include: very short focal lengths, large numerical apertures, small focal spot size, and low insertion losses [78]. The utility of these TIPE lenses has already been demonstrated via the production of an integrated acousto-optic Bragg modulator module on a LiNbO₃ substrate. Some preliminary testing utilizing the modulator for optical systolic array processing has also been undertaken [79]. These

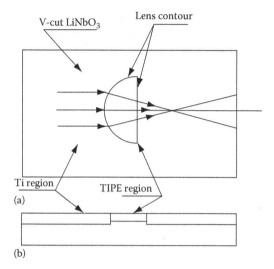

FIGURE 5.23
Planar waveguide lens in LiNbO formed by titanium-indiffused proton-exchange technique
(a) top side, (b) side view.

devices are expected to greatly facilitate the development and implementation of integrated and fiber-optic signal processing and computing.

5.10 Semiconductor Waveguide Fabrication Techniques

A wide variety of methods are used to fabricate optical waveguides and components. These include thin-film deposition, epitaxial growth, filled channel techniques, ion diffusion, and ion implantation. Several of these techniques have been discussed in other sections. This section covers the techniques most conducive to integration and production of optical circuits—namely, MOCVD, MBE, and ion implantation—along with a discussion on foundry facility utilization.

5.10.1 Ion Implantation

Ion implantation is a particularly valuable technique that offers a number of distinct advantages in comparison to the other fabrication methods [80]. In ion implantation a beam of atoms is ionized, accelerated to kinetic energies up to several million electron volts, and directed at a suitable material target. The resulting damage in the material and the introduction of foreign atoms can produce electronic, chemical and optical changes. The primary advantage of this technique lies in the precise external manner by which doping of materials can be controlled. Any known species can be introduced into a given material, and the dopant concentration is not limited by ordinary solubility considerations. (However, the concentration of dopant atoms occupying substitutional lattice sites is generally limited by the solid solubility.) The dosage can be monitored accurately and the depth profile of the implanted ions can be controlled by adjusting the implant energy. This permits realization of dopant profiles that could not be achieved by diffusion or any other technique. Also, implantation of materials can be performed over a wide range of temperatures, allowing unique interactions of the implanted ions. Precise device configuration is achievable due to the parallel development of high resolution GaAs lithography and masking, and through proper control and selection of the ion beam energy and fluence and the substrate doping level.

Ion implantation can produce a change in the refractive index of a crystalline semiconductor material through various physical mechanisms [81]. These mechanisms include damage to the crystal lattice, ultimately resulting in an amorphous region; introduction of dopant atoms into the lattice, causing a change in the polarizability of the unit cell; localized regions of stress in the lattice due to damage and the presence of a large number of dopant atoms in these regions; and the compensation of the free carriers in suitably doped materials. This last mechanism has been the primary method used to form optical waveguides in semiconductor materials through ion implantation. Of course, in any actual implementation the mechanisms are interrelated, and the change in the index of refraction of the implanted materials is due to the total effect of these processes. It is possible, however, to accentuate one of the mechanisms and to minimize the effects of the others. The most common ion species utilized for GaAs optical device fabrication is the proton (H+). These ions produce the least amount of damage to the crystal and minimize the contributions to the optical properties arising from the physical processes other than free-carrier compensation. Protons also are unlikely to occupy substitutional sites in the lattice and have the greatest projected range by direct energetic penetration without relying on any secondary or defect diffusion.

5.10.2 Ion-Implanted Semiconductor Annealing

As discussed in *Laser Annealing of Semiconductors* [82], the abrupt recrystalization process upon annealing ion-implanted silicon in the temperature range 500°C–600°C is in contrast to the more complex multistage process in the solid phase annealing of ion-implanted GaAs. Several publications [83] illustrate that damage removal of ion-implanted damage in GaAs occurs over a broad temperature range.

The nature of the solid phase annealing process was further revealed [84] as shown in Figure 5.24. Measured GaAs disorder, obtained from Rutherford

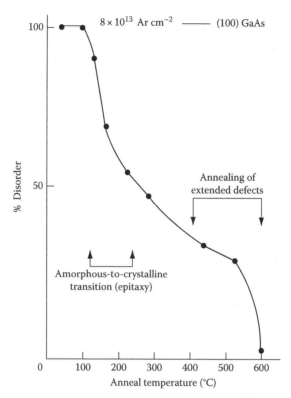

FIGURE 5.24

Normalized disorder, as measured by channeling, for 100 keV 8×10^{13} Ar cm^{-2}–implanted (100) GaAs plotted as a function of furnace annealing temperature (15 min anneals), indicating two annealing stages. (From Williams, J.S. and Harrison, H.B., In *Laser and Electron Beam Solid Interactions and Material Processing*, Gibbons, J.F. et al, eds., North-Holland Publishers, New York, 1981, p. 209.)

backscattering-channeling spectra, was plotted as a function of the anneal temperature following an amorphizing implant of 8×10^{13} Ar cm^{-2} at 100 keV. These data revealed two apparent annealing stages, the first occurring rather sharply in the temperature range 125°C–230°C (for 15 min anneals), followed by a second stage in the range of 400°C. Channeling cannot unequivocally reveal the nature of the residual disorder but, from observations of characteristic color changes during annealing, Williams and others [85] suggested that stage-1 recovery related to defects corresponding to amorphous regions in the crystal. Kular [86] correlated channeling, TEM, and electrical measurements, and indicated a third anneal stage in which the density of loops was progressively reduced.

Of further significance are measurements of electrical activity of ion-implanted GaAs that show continual improvement up to 9000°C [87]. In view of the anneal process described, the best electrical behavior is at least contingent upon the complete removal of crystal defects that arise during the low temperature annealing of implanted GaAs. This is not to suggest that complete removal of crystalline dislocation loops is sufficient to guarantee good electrical behavior. Other factors may certainly influence the electrical properties, following furnace annealing, such as point defect clusters and material impurities, particularly Cr distributions, and the nature of encapsulants used to protect the surface from decomposition.

Pulsed laser annealing of ion-implanted GaAs received considerable attention from several laboratories [88]. These studies indicated that uncapped wafers be laser processed in air at room temperature to provide good crystallinity, high substitutionality of implemented ions, and electrical activities superior to those achievable by conventional furnace annealing although, under certain conditions, decomposition of the surface was a problem.

Williams compared [89] the effectiveness of CW laser and furnace annealing in the removal of amorphous implant damage in GaAs, and found that both CW (argon ion) laser and the low-temperature furnace anneal produced good amorphous-to-crystalline recovery (stage-1 annealing). It became apparent, in both the pulsed and CW laser annealing experiments, that many of the problem areas associated with furnace processing of GaAs either remained as difficulties or gave way to significant problems associated with rapid heating and cooling effects [90]. Results obtained from both furnace annealing and the various transient annealing methods for implanted GaAs are summarized in Figure 5.25 [91]. Since generally only simple point defects are produced on implanted GaAs, with no extended defects or loops, only first-stage annealing (125°C–230°C) provides an effect, which is advantageous to optical performance.

5.10.3 MOCVD: Growth and Evaluation

MOCVD and MBE are discussed and contrasted in an earlier section on material growth. Recent improvements in crystal growth capabilities and heteroepitaxial deposition of very low defect density structures, using either MOCVD or MBE, have significantly advanced the optical circuit device producibility and performance factors.

Figure 5.26 illustrates the operation of the vertical atmospheric-pressure MOCVD reactor [92]. In this technique, "vapor-phase mixtures containing the metalorganic compounds are pyrolyzed at or near the surface of a heated substrate, where they combine to form the deposited layer. Most sources are liquid near room temperature with relatively high vapor pressures, and so may be carried to the reaction zone by bubbling a carrier gas such as high purity hydrogen through the liquid source" [93]. As an example, GaAs is usually grown using trimethylgallium plus pure arsine gas producing GaAs and carbon tetrahydride (CH_4) by-product. "Deposition proceeds as reactants decompose in a stagnant boundary layer just above the surface of the wafers." Commercial equipment is capable of processing up to twelve 3 in.

	Furnace	Continuous-wave laser/electron beam	Pulsed laser/electron beam
Anneal mechanism	Solid phase	Solid phase	Liquid phase
Optimum anneal conditions	>850°C for best electrical properties	one to 5S dwell with substrate heating (laser) and raster scan (electron beam)	Narrow energy window at energies just sufficient to melt below damage layer
Decomposition	Severe at >600°C: use cap or other preventative measures	Surface oxidation in air: use inert ambient or vacuum	Significant decomposition only at high powers and pulse lengths: surface damage at lower powers
Microstructure	Always some extended defects	Not yet examined	Extended defect-free but often some surface damage: likely quenched in point defects
Surface topography	Featureless	Featureless (optimum conditions): slip and cracking with excessive thermal gradients	Featureless (optimum conditions): rough surface at high power
Electrical properties	Good activity: always below 10^{19} cm^{-3} for n type	Limited data but good activity for low to medium dose n-type: poor results for non optimized anneals: mobilities close to theory, where available	High activity (>10^{19} cm^{-3}) for both n- and p-type: activity decreases on subsequent furnace annealing: cannot activate low-dose n-type: mobilities anomalously low
Implant redistribution	Diffusion broadening a problem with some dopants (e.g., Zn. S) at high temperatures	Few data exist: probably no redistribution	Little redistribution for low power just above threshold: dopant redistribution and "zone refining" can occur in melt
Substitutional solubility (to GaAs)	Never exceeds 5×10^{19} cm^{-3}	Not measured	Exceeds equilibrium value (>10^{20} cm^{-3}) not good correlation with electrical activity

FIGURE 5.25
Comparison of the various annealing methods for ion-implanted GaAs.

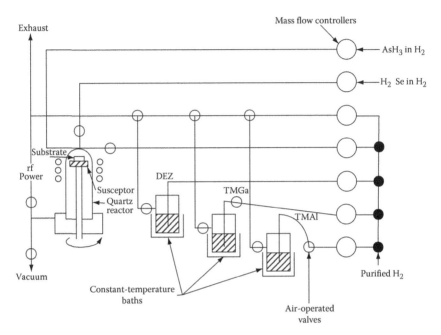

FIGURE 5.26
Vertical atmospheric-pressure MOCVD reactor.

wafers in a single processing cycle and up to three processing cycles may be completed in a production shift.

The aluminum composition (x) of the alloy is determined from the peak energy of the band-edge emission of photoluminescence spectra for single layers. A special facility can profile the photovoltage of a multilayer structure at the same time that a carrier profile is obtained by electrolytic removal of material in conjunction with computer controlled capacitance-voltage determinations. Deep level traps in the material are investigated by the Deep Level Trap Spectroscopy (DLTS) technique and other related measurements.

5.10.4 MBE: Growth and Evaluation

Molecular beam epitaxy, which involves the reaction of one or more thermal beams of the desired atoms or molecules with a crystalline substrate under ultrahigh vacuum conditions (10^{-10}–10^{-11} Torr background, 10^{-6} Torr operating pressure), can achieve precise control in chemical compositions, doping profiles, and layer thickness. A typical MBE system for growth of GaAlAs is shown in Figure 5.27. Uniformity of chemical composition and doping profile are typically better than 1% with thickness variations less than +0.5%. An additional advantage is that no high temperature processes are required, so that very minimal bulk diffusion effects occur. Single-crystal multilayer AlGaAs atomic layers can be grown by MBE. Atomically smooth growth

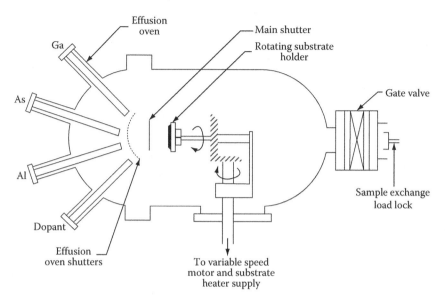

FIGURE 5.27
A typical MBE system for growth of AlGaAs.

surfaces allow complex doping and compositional structures and synthesis of artificial materials with prescribed characteristics.

Characterization of MBE samples occurs first during growth with the in situ capability of RHEED and Auger electron spectroscopy which provide information for growth conditions, allowing preparation of high quality samples. After growth of samples, x-ray diffraction and electron microscopy provide information on crystal structure, crystal orientation, and crystal perfection. Transport measurements provide information concerning carrier concentration and mobility. Raman spectroscopy can provide information regarding the quality of interfaces.

Complex optical circuit structures require epitaxial layer uniformity and precision in thickness, doping, and AlGaAs composition. Depending upon the particular design, there are advantages to both MOCVD and MBE. Tables at Figures 5.28 and 5.29 compare material properties, parameter controls, and system cost for the two techniques.

5.10.5 MBE Development in Space

Past and present experience in submicron optoelectronic and electronic devices employing single crystal GaAs in epitaxial interface with single crystal AlGaAs has clearly indicated that the limiting criteria for optimal device performance is the purity of the AlGaAs. Even under the best starting condition vacuums available on earth (10^{-11}–10^{-12} Torr), there exist serious defects caused by impurity atoms and by water vapor as well as Co_2 and O_2. Such

	MOCVD	MBE
Flexibility	Most compounds and alloys	Difficulty with phosphorus and Alloys with two column V elements
Purity		
GaAs	μ_{77} 139,000	μ_{77} 125,000
InGaAs	μ_{300} 11,800	μ_{300} 8,800
InP	μ_{77} 58,900	μ_{77} 34,000
Photoluminescence		
AlGaAs	Comparable to LPE	Less than LPE
Others	Comparable to LPE	Comparable to LPE
Uniformity		
Doping	2.5%	2.5%
Thickness	2.5%	1.0%
Composition	0.005	0.005
Interfaces		
AlGaAs/GaAs	1–2 Monolayer	1 Monolayer
InAlAs/InGaAs	<50 A	—
InP/InGaAs	—	1 Monolayer
Defect density	Dislocation Density	$1–3 \times 10^3/cm^2$ oval defects

FIGURE 5.28
MOCVD versus MBE—Comparison of Material Properties. (From Dapkus, P.D., *J. Crystal Growth*, 68, 345–355, 1984.) [94]

	MOCVD	MBE
Growth rate and/or composition	Source temperature (±0.25°C)	Source temperature (±0.4°C)
	Flow rates (±1%)	Substrate temperature (±5°C)
	Substrate temperature (±5°C)	
Doping	Flow rates	Source temperature
	Substrate temperature	Substrate temperature
Purity	Source purity	Vacuum materials
	Growth temperature	Source purity
Interface abruptness	Reactor design, flow rates, substrate	Shutter speed
		Rotation speed
		Substrate temperature
System cost	$200–$400 K	$700–$1000 K
Production throughput	≤900 cm²/run	≤20 cm²/run
Safety considerations	Poisonous materials	Toxic wastes
	Pyrophoric materials	
	Toxic wastes	

FIGURE 5.29
MOCVD versus MBE—Comparison of parameter controls and system cost. (From Dapkus, P.D., *J. Crystal Growth*, 68, 345–355, 1984.) [94]

impurities adversely affect the energy gap of the AlGAAs and the epitaxial match lattice parameters for single crystal growth and interface requirements, as well as adversely altering doping carrier concentrations and establishing unwanted localized or delocalized energy band states.

Growth of optoelectronic waveguides and submicron devices in the ultra-high vacuum conditions (10^{-14} Torr) behind the wake shield of the NASA Space Shuttle (a working volume of about $50\,\mathrm{m}^3$) or a space-based manufacturing facility will greatly enhance device performance. Improved performance factors include superior confinement of the exciton cloud in the layer of lower E_g for ultra-fast switching and modulation of optical quantum well devices; better homogeneity of layers and waveguides, and constancy in the mismatch of refractive indices throughout the waveguide boundary interface of an optical coupler of interferometric device; and improved gain characteristics of high electron mobility transistors.

5.11 GaAs Foundry Capabilities

About 25 years ago the first GaAs ICs were introduced in the form of microwave prescalers. A parallel may be seen between the development of silicon based industries and the growth pattern of the GaAs industry. First came the transistors, then simple logic applications such as flip-flops, and from this arose full multifunction monolithic capability. At this point, high quality GaAs substrate material is being achieved and several foundries are in mass production of MMICs for the microwave industry. A major application of this MMIC technology centers on the need for low cost highly reliable microwave components for Phased Array Radar Transmit/Receive Modules. Optical applications of MMIC technology are discussed in a later section. MMIC processing and foundry capabilities are directly applicable to the requirements and problems of optical device fabrication.

Several wide bandwidth functional building blocks are available for design use, including switches, phase shifters, and medium performance amplifiers, with the GASFET being the key building block. Applications and the GASFET technology are discussed further in a later section. The designer will save valuable development time and money by adjusting the circuit architecture to take advantage of available functional blocks. While the discrete GASFET manufacturers are producing devices for low noise applications with gate lengths of less than $0.3\,\mu\mathrm{m}$. MMIC foundries are capable of routinely producing only $0.5\,\mu\mathrm{m}$ FETs, restricting the microwave designer to a "medium low noise amplifier" (implying a noise figure of $3\,\mathrm{dB}$ at $12\,\mathrm{GHz}$, for example). Heterostructure devices, such as the high electron mobility transistor (HEMT), are currently in production at MMIC foundries.

Many foundries still produce three inch wafers as standard. The hope is that eventually the growth of high purity GaAs in larger diameter slugs will help raise processing yields and lower the cost of MMIC devices. One way in which MMIC manufacturers are increasing their yield percentages is by the use of process control monitors (PCMs) on the wafers. By using PCMs in conjunction with specialized test equipment that has been developed specifically for MMIC production purposes, manufacturers are able to pull bad wafers from the production line before too much time and effort has been spent on processing a wafer that will not yield a profitable amount of useable MMICs. Electrostatic discharge of the GaAs devices is accomplished by integrating shunt diodes onto the die at the device inputs and outputs.

In designing an optical circuit employing monolithic microwave devices, the devices used in the design must be within the standards set by the foundry chosen to manufacture the chips. The use of computer-aided design (CAD) and computer-aided engineering (CAE) tools for the MMIC portion of the system usually produces significant cost efficiency. Fraser and Rode of Triquint Semiconductor have published a list of design strategies they feel will provide a firm basis for efficient and reliable MMIC designs. A few of the major points from the list follow [95]:

1. MMIC costs are not based on the number of FETs used in the design, so they should be used freely.
2. MMIC is best suited for medium performance functions; circuit complexity is inversely proportional to the yield.
3. Start with simple designs to prove concepts and then move on to more complex topologies.
4. The use of lumped elements below 18 GHz will reduce circuit size without sacrificing performance.
5. Make use of the standard cell libraries provided by the foundry to avoid reinventing the wheel.
6. Design for testability by providing ample test points at easy to access places on the wafer.
7. During computer optimizations of the design it is imperative that the effects of process variations and material spreads are included.

A typical MMIC design sequence from performance definition through the foundry layout review is as follows [96]:

1. Define MMIC performance requirements.
2. Select the basic circuit and layout topology.
3. Simulate circuit performance using CAE techniques.
4. Analyze the design over temperature and process variations.

5. Optimize the design and do a rough circuit layout.

6. Calculate the device parasitics and incorporate them into the design.

7. Re-optimize the design over temperature and process variations.

8. Run the circuit through a design rule checker to ensure that none of the basic layout rules have been violated.

9. Conduct a layout review at the foundry.

Computer analysis and synthesis software has been developed to meet the needs of the microwave designer by a number of sources. Companies are now providing the software packages that enable the engineer to design circuitry, perform analysis and optimization functions, and transfer layouts. In addition, they furnish the user with foundry libraries to enable simulation of a MMIC implementation of the user's circuitry. Figure 5.30 [97] indicates the range of CAE/CAD software available from one manufacturer and demonstrates how this software fits into a systematic design flow.

After the design sequence is completed the foundry is generally assigned the task of generating the masks required to produce the finished wafer. After the wafer is produced, the foundry test facility is able to run tests on the various dies on the wafer according to test points that have been made available by the designer. A typical test setup [98] for wafer testing is shown in Figure 5.31.

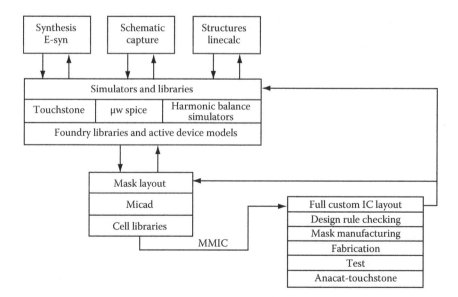

FIGURE 5.30
EESOF product suite. (After Parrish, P. of EESOF, Computer-aided engineering in GaAs monolithic technology, *Monolithic Technology*, October 1987.)

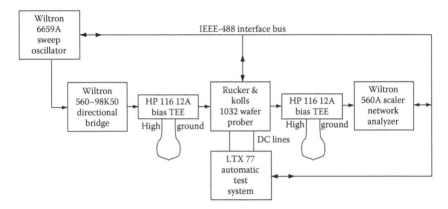

FIGURE 5.31
Typical wafer mapping test station. (After Dammann, C. et al., Microwave die sort and wafer mapping for GaAs MMIC manufacturing, *Monlithic Technology*, October 1987.)

FIGURE 5.32
Typical wafer test station display. @=high current, #=low current, %=bad rf, &=low gain, *=high refl, $=passed.

With this configuration, the operator may run a variety of DC and RF tests which generate a computer display similar to that [99] depicted in Figure 5.32. With this information stored in the system controller, the good dies may be scribed and separated from the unusable. If this prototype hardware passes testing, the design may go to production. If the design is flawed, then the design layout, mask generation, and prototype manufacturing cycle must be repeated.

The cost of a single foundry prototype run, including mask generation, is generally between thirty and fifty thousand dollars depending upon the process required [100]. Although this is expensive, twenty to thirty designs can be integrated on one mask so that the cost per MMIC design may be as low as two to three thousand dollars. Once in production, the cost per wafer will drop by more than an order of magnitude. A failure rate of one GaAs FET per 2×10^6 h may be expected for newer medium-integration products [101]. This rate is lower than the failure rate for discrete GASFETs. This result is in accordance with trends established for silicon ICs [102]. At 125°C channel temperature, and with current densities within the limits of the design rule checker, the failure rate of a standard MMIC design should be less than 0.01% per 1000 h [103].

As technical advances in this arena continue, submicron capability becomes cost effective. Submicron gate length FETs promise lower noise figures and increased sensitivity of microwave receivers. By increasing the standard wafer size and tightening manufacturing process controls, higher yields per wafer and lowered MMIC cost are realized. Integration of HEMT processing into the MMIC arena aids in the expansion of MMICs into the millimeter wave frequency band.

5.12 Emerging Commercial Devices and Applications

Dozens of companies worldwide are currently offering integrated optical circuits and components. Many operate pilot production lines for a wide range of components. (Use of pilot production or foundry facility for development of custom OICs was discussed in an earlier section.) Many of these companies are already well established in the related materials, optical communications, and defense system technologies. The components in the market target a wide range of applications, including the following:

- Telecommunications (line-of-sight and fiber optic)
- Optical computing
- Laser range finding
- Optical proximity fusing
- Fiber sensors
- Target designation and tracking
- Industrial measurement and control
- Optical interconnects
- Medical instrumentation

In particular, currently available components perform various critical functions in fiber-optic communication and sensing systems, such as modulation (phase and intensity), switching, mode conversion, splitting/coupling, and signal processing. Many of the companies offering integrated optic devices concentrate on the LiNbO$_3$ and glass technologies, but a wide variety of silicon and compound semiconductor devices are available as well.

Because of the range and diversity of applications for integrated optics, estimates of the markets are difficult to project. Many companies and R&D labs have custom capability but do not advertise. Companies typically advertise a year or two ahead of availability. One set of predictions [116] estimated a total integrated optical circuit market of $5M in 1987, growing to $100M in 1990, and to $500M by 1995. Eighty percent of the 1990 market was predicted to be LiNbO$_3$ devices, but 50% of the 1995 market would involve compound semiconductor devices. The applications are predominantly in the fiber sensor, communication, switching, biomedical, and instrumentation areas.

The integrated optical product data shown in Figure 5.33 through 5.41 was supplied by a number of the companies initially offering commercial devices 20 years ago. These are representative of the technology available from a number of sources. Direct performance comparisons are not made, since specific applications will determine the most suitable device. Specific device characteristics should be obtained from manufacturers.

Future systems requirements are as diverse as the applications to be addressed. Critical components include tunable lasers, optical routers and

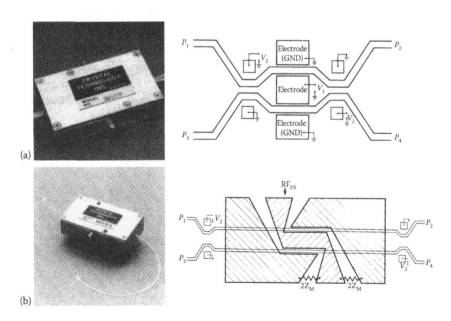

FIGURE 5.33
(a) LiNbO$_3$ 2 × 2 switch, (b) LiNbO$_3$ traveling wave 2 × 2 switch.

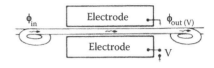

- dc to 3 GHz
- 850, 1300, or 1550 nm
- 5–15 V drive voltages

- 7–9 V drive voltages
- 5 GHz BW
- Drift free dc bias operation

FIGURE 5.34
LiNbO$_3$ phase modulators.

high density crossconnects, wavelength converter devices, optical switches, add-drop multiplexers, amplifiers and transceivers, fiber gratings, equalizing filters, bandwidth management, circulators, modules, and integrated packages, and a range of interface components and connectors. Successful developers require flexible material systems and technology approaches. Integration of key functions on planar optical platforms improves reliability and lowers ultimate integration costs. Mixed optic and electronic solutions best adapted to market needs provide the optimum strategy. Fiber grating devices providing optical networking enhancements include switches, adjustable filters, network monitors, add-drop filters, gain flatteners, and dispersion compensators.

5.12.1 Fiber-Optic Couplers

The most successful coupler manufacturing technology to date is the fused biconical taper (FBT) process. This involves placing two center-stripped fibers

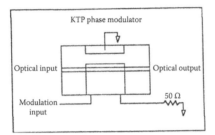

- Fiber pigtailed
- Operation to 15 GHz
- Wide range of wavelengths (0.4–1.6 µm)

(a)

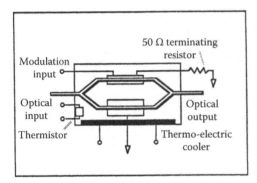

Lithium niobate modulator

- Mach–Zehnder modulator/switch
- 1300 or 1550 nm operation
- Operation to 4 GHz
- 20 dB extinction ratio
- Fiber pigtailed

(b)

FIGURE 5.35
(a) KTP phase modulator; (b) LiNbO$_3$ Mach–Zehnder intensity modulators (Courtesy of British Telecom & Dupont [BT&D Technologies], London, U.K.);

close together in a fixture and heat stretching the glass over a hydrogen-oxygen flame until a fused region is created. This is performed while monitoring the optical power through the fibers until a desired optical performance is achieved. The fused region is then packaged in a stable substrate.

Technical and sales leaders included Gould, ADC, Corning, FBT, E-Tek, Aster, IOT, IOC, and AMP. Gould is the inventor of the FBT process and holds some of the key patents. Manufacturing was transferred to Swiftlink in India, who supplies dual window (1310/1550) FBT components. Corning patented a multiple index coupler, where a circular glass substrate is applied around the fibers during fusion. This is claimed to be more stable, but is more complex. E-Tek designed and developed an automated coupler workstation, with excellent repeatability and polarization dependent loss.

- 5–7 V drive voltages
- 5 GHz bandwidth
- Drift free dc bias operation

(c)

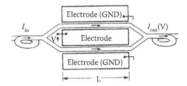

- dc to 3 GHz
- 4–11 V drive voltages
(d) • 850, 1300, or 1550 nm

FIGURE 5.35 (continued)
(c) (Courtesy of Pilkington Guided Wave Optics & Barr & Stroud, Ltd., Glasgow, U.K.);
(d) (Courtesy of Crystal Technology, A Siemans Company, Palo Alto, CA.)

The manufacturing technology for FBT is quite challenging. FBT devices are relatively large and require significant packaging design for stable operation. At best, this technology can support six fiber ports with reasonable port-to-port uniformity and optical performance. Port counts of higher than six require fusion splicing and bulky enclosures. FBT devices require dedicated workstations and groups of operators fabricating one fused region at a time although recent efforts to automate the process show promise. Average daily yields often do not exceed 70%.

5.12.2 Performance Testing Issues for Splitters

The optical parameters most often tested are the following:

- Split ratio—this is the percent of optical power measured at an output fiber compared with the total output power. This is also measured as insertion loss in dB. Most splitters are 50/50 equal split.

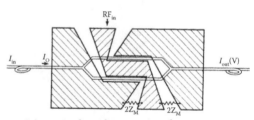

Schematic of traveling wave interferometer

● dc to 10 GHz
● 4–11 V drive voltages
● 850, 1300, or 1550 nm

FIGURE 5.36
Traveling-wave intensity modulator. (Courtesy of Crystal Technology, A Siemens Company, Palo Alto, CA.)

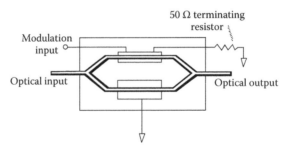

● Fiber pigtailed
● Operation to 15 GHz
● Wide range of wavelengths (0.4–1.6 μm)

FIGURE 5.37
KTP Mach–Zehnder modulator.

- Excess loss—the amount of optical power lost in a device, in addition to the insertion loss.
- Passband—the wavelengths at which the performance specifications apply. Splitters are generally highly wavelength sensitive.
- Directivity—the amount of power reflected relative to the input power to a device, measured exiting an input fiber. FBTs typically exhibit better than 60 dB directivity.
- Isolation—when a splitter is used as a demultiplexer, isolation is a measure of the coupler's ability to separate two or more wavelengths. Isolation is usually measured as a dB difference in the measured optical power of one wavelength compared to another.

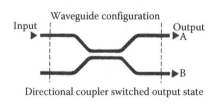

Directional coupler switched output state

- 0.85, 1.32, and 1.52 µm
- 15 V switching voltages
- dc to >100 MHz frequency response

(a)

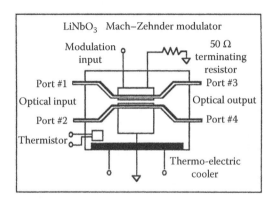

- 6 V switching voltage
- Operation to 4 GH$_Z$
- Fiber pigtailed
- 12 dB channel separation

(b)

FIGURE 5.38
LiNbO$_3$ directional couplers modulators. (a) (Courtesy of Pilkington Guided Wave Optics);
(b) (Courtesy of Barr & Stroud, Ltd., Glasgow, U.K.)

- Uniformity—most often used with more than two fiber I/O ports, uniformity is the maximum difference in insertion loss in dB from one port to another.

- Polarization—also designated polarization dependent loss (PDL) in splitters. The optical power transmission of fiber-based devices is often dependent on the orientation of the electromagnetic wave entering the device. PDL is a measure of how insertion loss changes as the orientations or states of polarization (SOP) change. In most quality devices, PDL measures less than 0.2 dB.

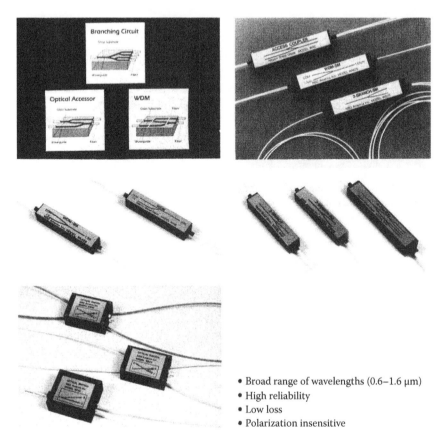

• Broad range of wavelengths (0.6–1.6 μm)
• High reliability
• Low loss
• Polarization insensitive

FIGURE 5.39
Ion-exchange borosilicate glass waveguide single mode fiber-optic contents.

The next step in device complexity beyond the splitter is the single mode tree or star coupler. These can be $1 \times N$, $2 \times N$, $4 \times N$, etc. These components usually have equal split ratios but can be manufactured with asymmetric ratios. Dual wavelength devices are more complex and costly. Port counts greater than four usually require splicing of two or more devices and large enclosures or ultimately integration in photonic circuits.

Multimode trees and stars are similar to the single mode devices, but rarely fabricated with asymmetric split ratios. Constructed with fibers from $50/125$ to $200/230\,\mu m$, they are generally used in high power industrial and medical sensing while the single mode devices are used more in the CATV and DWDM markets. For example, 200/230 devices are used to couple signals from intravenous catheters in arterial blood-gas sensor systems. 100/140 devices are used to link pressure catheters to intracranial pressure monitors. They are also used in rpm monitors for steam turbines.

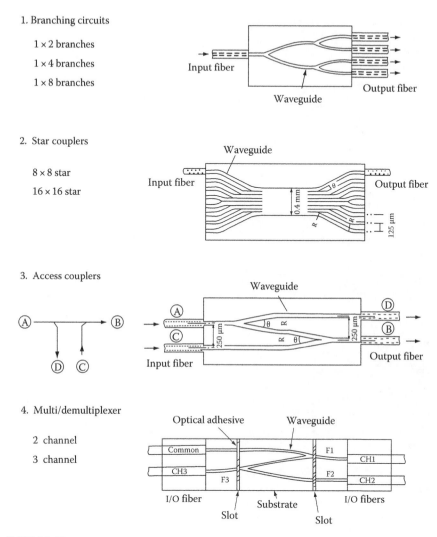

1. Branching circuits

    1 × 2 branches

    1 × 4 branches

    1 × 8 branches

2. Star couplers

    8 × 8 star

    16 × 16 star

3. Access couplers

4. Multi/demultiplexer

    2 channel

    3 channel

FIGURE 5.40
Applications of graded-index waveguides.

Wavelength division multiplexers/demultiplexers are manufactured using the FBT process for sharp tapers, exploiting interference effects between guided wave devices. These are usually 1310/1550 or 980/1550 devices. Typical isolation in a 1310/1550 device is 20 dB. High isolation devices (>40 dB) can be constructed by splicing two or more devices in series.

Low PDL FBT tap couplers with asymmetric split ratios, typically 1/99 or 5/95, are used in feedback control systems to monitor laser output of sensitive optical instruments where the state of polarization can affect system

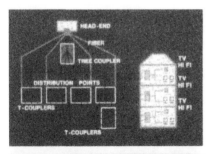

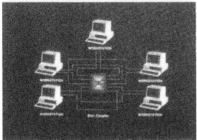

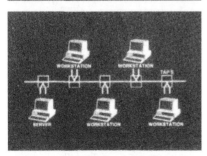

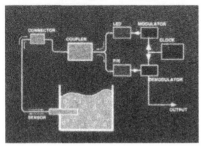

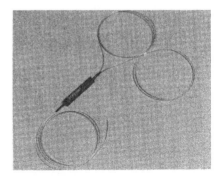

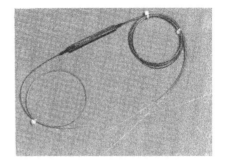

- Splitters
- Star couplers
- Wavelength division multiplexers
- Optical power taps

- Low loss
- High directivity
- Uniform power splitting
- Insensitive to modal dependencies
- Wide operating wavelength range
- Consistent optical, mechanical, and environmental performance

FIGURE 5.41
Ion-exchange glass technology for fiber-optic components.

performance. Low reflection terminations (LRT) are manufactured by creating an asymmetric sphere or fiber loop to prevent reflections from unused splitter ports in bi-directional systems causing unstable laser performance or noise.

5.12.3 Passive Optical Interconnects

Passive optical components guide, filter, route, adjust and stabilize optical signals transmitted through an optical network. Recent technological advances in the design and production of both passive and active components allow communication equipment suppliers to optimize fiber-optic systems. The ongoing trend is to integrate increasing optical component count into compact modules and custom solutions.

The concept of flexible optical circuits originated at Bell Labs in the late 1980s [104–106]. Originally named "Optiflex" the concept is now produced in similar fashion by nearly a dozen companies including FCI, Polyguide Photonics, Tyco, AIT, Molex, NTT, USConec, and others. Fiber crossovers are designed to minimize stress in the fibers. Several companies manufacture multifiber array connectors suitable for use with the flexible optical routing circuits. These employ precision molded ferrules as well as micromachined silicon v-grooves for accurate alignment of fibers for optimal coupling configurations. Complex fiber management schemes developed at DuPont and later licensed to PolyGuide Photonics, aka Optical Crosslinks Inc., provide the capability to manufacture flexible routing circuits as well as related polymer waveguide based passive components [107].

One key requirement for optical communication systems is the perfect optical shuffle or optical flex board shown in Figure 5.42. Eight sets of eight-input fiber arrays are routed to each of eight sets of eight-fiber output arrays. In other words, one fiber from each of the inputs goes to one fiber of each of the outputs. In Figure 5.42, 1:8 splitters are used to create the eight sets of eight-fiber inputs; and an 8×1 switch is used to direct each of the output fibers to a set of single-fiber outputs. While the specific optical wiring may vary from one system application to another the basic optical shuffle component provides the fundamental building block.

This can be accomplished as shown in Figures 5.43 and 5.44, with various fiber routing and packaging schemes. In Figure 5.43, the fibers are ribbonized into 12 sets of 12-fiber arrays, and one fiber from each of the arrays goes to one of the output arrays. The fiber crossovers are housed in a "black box" called the shuffle. A variant of this scheme is shown in Figure 5.44, where the ribbonized fibers enter a sheet of material which contains the routing paths for the fibers which are shuffles as before. The ribbonized fibers are terminated in these examples with optical connectors called the MPO. This approach will work with the variety of mulitfiber connectors available, for example, MTP, MPX, MAC, miniMAC, etc., as the fibers are always ribbonized with 250 μm center-to-center spacing. In the assembly process, standard

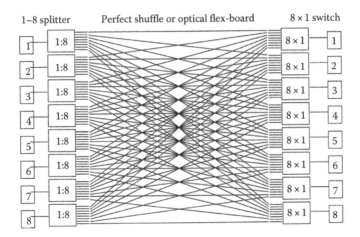

FIGURE 5.42
8×8 optical switch component.

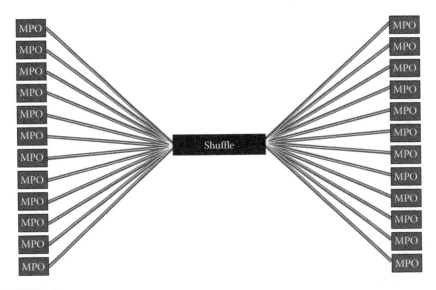

FIGURE 5.43
Optical shuffle.

communication grade optical fiber is laid onto an adhesive substrate material such as Kapton. Following the routing process, a thermoplastic material is laminated over the optical circuit to encapsulate and protect the circuit during handling and environmental extremes.

Another application for optical routing circuits is shown in Figures 5.45 and 5.46. In these cases the optical circuit provides interconnection of optoelectronic components on a circuit card, such as optical transceivers and

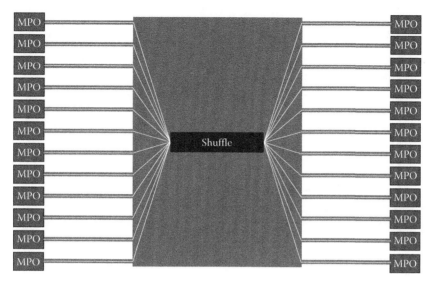

FIGURE 5.44
Optical flex circuit with perfect shuffle function.

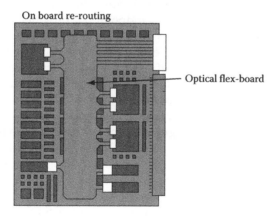

FIGURE 5.45
Optical routing device on circuit board.

processors, as well as a connection to the card edge or motherboard, and to a routing cabinet. Other applications include crossovers, splitters, couplers, and multiple simplex or multi-fiber termination points. Routing circuits are tailored for each system application. In most cases the only limitations on the fiber routing paths are the minimum radii of curvature allowable for the individual fibers. Encapsulation protects the fibers from debris, moisture, and handling. This provides maximum flexibility to the equipment system designer.

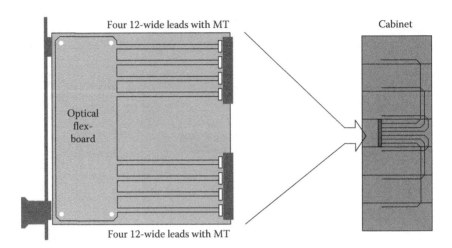

FIGURE 5.46
Optical configuration board.

5.12.4 Configuration of a Curved Transition Waveguide

Intersecting waveguides may be joined either abruptly or smoothly. Abrupt joining is easier from a fabrication standpoint but provides undesirable discontinuities to the traveling waves. The ideal coupling is a curved section, joined tangentially to the end of each individual waveguide. To fabricate such a section, one needs to specify the radius of curvature of this section and one needs to know the location of the center of this curvature relative to some well-established reference point.

Figure 5.47 shows the intersection of waveguides at the Double-Y output. In this case, a single waveguide is split abruptly at point D, into two waveguides which separate at an angle β. This abrupt separation is used so that the angle where the split occurs remains finite. It is desirable to keep this angle small to permit good coupling between the guides. On the other hand, one also desires a larger spread between the guides, in a reasonable distance, than would be attainable with the angle β. So the angle is further increased by the angle γ to a total of α. The following derivation yields the necessary values to locate a curved section of waveguide between points A and B.

The problem can be specified in terms of the angles α and β and the distances a and R. The segment $\overline{DE}$ is the distance a between the intersections of the individual guides with the straight output guide (inclined at $0°$), and R is the radius of curvature of the connecting section measured from point C. Other angles in the figure are related as follows:

$$\gamma = \alpha - \beta \tag{5.32}$$

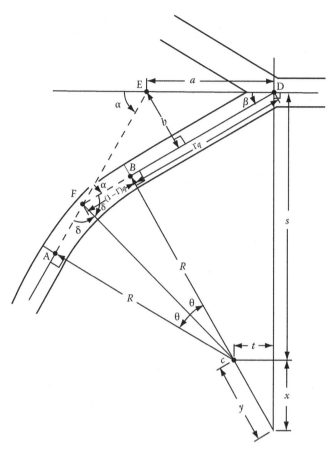

FIGURE 5.47
Geometry of curved transition waveguide.

$$\theta = \frac{\gamma}{2}$$

$$\delta = 90 - \theta$$

The distance from the waveguide split to the imaginary intersection of the two angled waveguides, the segment $\overline{DF}$, is defined by

$$q = \alpha \left(\cos\beta + \frac{\sin\beta}{\tan\gamma} \right) \tag{5.33}$$

It is seen that point B, the beginning of the curved section, divides q into two sections, and Γ represents the percentage of this division. The radius of curvature is found to be

$$R = \frac{(1-\Gamma)q}{\tan\theta} \tag{5.34}$$

$$\Gamma = 1 - \frac{R}{q}\tan\theta$$

The center of curvature C is located at distances t and s relative to point D. In order to find t and s we first find x and y

$$x + R = \frac{\Gamma q}{\tan\beta}$$

or

$$x = \frac{\Gamma q}{\tan\beta} - \frac{(1-\Gamma)q}{\tan\theta}$$

and

$$y = x\cos\beta$$

or

$$y = \Gamma q\frac{\cos\beta}{\tan\theta} - (1-\Gamma)q\frac{\cos\beta}{\tan\theta}$$

then

$$t = x\sin\beta$$

or

$$t = q\left[\Gamma\cos\beta - (1-\Gamma)\frac{\sin\beta}{\tan\theta}\right]$$

and

$$\delta = \frac{\Gamma q}{\sin\beta} - y$$

or

$$\delta = q\left[\Gamma\sin\beta + (1-\Gamma)\frac{\cos\beta}{\tan\theta}\right]$$

A final parameter of interest is the arc length of the curved segment:

$$\mathrm{Arc}\overline{AB} = 2R\theta \tag{5.35}$$

In summary, point B, the start of the curved section, is found by starting at point C, going a distance R in the $-y$-direction, and rotating an angle β. The length of the curve to point A is made by additional rotation of 2θ.

References

1. Hungsperger, R. G. 1984. *Integrated Optics: Theory and Technology*, 2nd edn. New York: Springer-Verlag.
2. Maractili, E. A. 1969. Dielectric waveguide and directional coupler for integrated optics. *Bell System Tech. J.* 48(7):2102.
   Goell, J. E. 1969. A circular-harmonic computer analysis of rectangular dielectric waveguides. *Bell Syst. Tech. J.* 48(7):2133–2160.
   Schlosser, W. and H. G. Unger. 1966. Partially filled waveguides and surface waveguides of rectangular cross section. *Advances in Microwaves*. New York: Academic Press, pp. 319–387.
   Bartling, J. Q. 1969. Propagation of electromagnetic wave in an infinite rectangular doelectric waveguide. *J. Franklin Inst.* 287(5):389–407.
   Shaw, C. B., B. T. French, and C. Warner III. Further research on optical transmission lines. *Scientific Report* (2) Contract AF449(638)-1504 AD 625 501. *Autonetics Report* (C7-929/501), pp. 13–44.
3. Barnoski, M. K., ed. 1974. *Introduction to Integrated Optics*. New York: Plenum Press, pp. 101–102.
4. Bradley, J. C. and A. L. Kellner. 1983. Multimode structure of diffused slab channel waveguides in uniaxial crystals. *SPIE East 83*, April, Arlington, VA.
5. Marcatili, E. A. 1969. Dielectric waveguide and directional coupler for integrated optics. BSTI 48(7): 2071–2102.
6. Kuznetsov, M. September 1983. Expressions for the coupling coefficient of a rectangular waveguide directional coupler. *Opt. Lett.* 8:499 ff.
7. Hutcheson, Lynn. June 1986. Private conversation with author.
8. Mentzer, M. A. 1986. Ion implantation fabrication of gallium arsenide integrated optical components. PhD dissertation, University of Delaware.
9. Mentzer, M. A., R. G. Hunsperger, J. Bartko, J. M. Zavada, and H. A. Jenkinson. 1984. Ion implanted GaAs integrated optics fabrication technology. *Proceedings of SPIE* 517.

10. Mentzer, M. A., R. G. Hunsperger, S. Sriram, J. Bartko, M. S. Wlodawski, J. M. Zavada, and H. A. Jenkinson. 1985. Ion implanted optical waveguides in gallium arsenide. *Opt. Eng.* 24:225–229.
11. Boyd, J. T., R. W. Wu, D. E. Zelmon, A. Nauman, and H. A. Timlin. 1984. Planar and channel optical waveguides utilizing silicon technology. *Proceedings of SPIE* 517.
12. Webb, R. P., and W. J. Devlin. 1986. Resonant semiconductor-laser amplifier as an optically controlled wavelength selector. *Conference on Lasers and Electro-Optics, Technical Digest*, June, p. 354.
13. Jewell, J. L. 1986. Semiconductor etalons for optical information processing. *Conference on Lasers and Electro-Optics, Technical Digest*, June.
14. Lee, Y. W., H. M. Gibbs, J. L. Jewell, J. F., Duffy, T. Venkatesan, A. C. Gossard, W. Wiegmann, and J. H. English. 1986. Speed and effectiveness of window-less GaAs etalons as optical logic gates. *Conference on Lasers and Electro-Optics, Technical Digest*, June.
15. Jewell, J. L., Y. H. Lee, H. M. Gibbs, N. Peyghambarian, A. C. Gossard, and W. Wiegman. 1985. The use of a single nonlinear fabry-perot etalon as optical logic gates. *Appl. Phys. Lett.* 46:918.
16. Sturge, M. D. 1962. *Phys. Rev.* 127:763.
17. Stoll, H., A. Yariv, R. G. Hunsperger, and E. Garmire. 1974. Proton-implanted waveguides and integrated optical detectors in GaAs. *Digest of Technical Papers.* OSA topical meeting on integrated optics, January, New Orleans, LA, pp. 21–23.
18. Tracey, J. C., W. Wiegman, R. Logan, and F. Reinhart. 1973. Three-dimensional light guides in single-crystal GaAs/GaAlAs. *Appl. Phys. Lett.* 22:10.
19. Tien, P. K. 1971. Lightwaves in thin films and integrated optics. *Appl. Opt.* 10(11): 2395–2413.
20. Blum, F. A., D. W. Shaw, and W. C. Holton. 1974. Optical striplines for optical integrated circuits in epitaxial GaAs. *Appl. Phys. Lett.* 25:116.
21. Tien, P. K. Lightwaves in thin films and integrated optics.
22. See, for example, Hunsperger, R. G. 1985. *Integrated Optics: Theory and Technology*, 2nd edn. Heildelberg: Springer Verlag.
23. Goell, J. E. 1974. Loss mechanisms in dielectric waveguides. *Introduction to Integrated Optics*, Barnoski, M. K., ed. New York: Plenum.
24. See, for example, 1984. *Proceedings of the Second International Conference of Metalorganic Vapor Phase Epitaxy*, April, Sheffield, U.K.
25. See, for example, 1986. *Abstracts of the Third International Conference on Metalorganic Vapor Phase Epitaxy*, April, Universal City, CA.
26. Hutcheson, L. D. 1986. AlGaAs/GaAs processing and growth compatibility for monolithic integration. *Proceedings of SPIE* 612, January.
27. Chen, T. R., L. C. Chiu, K. L. Yu, U. Koren, A. Hasson, S. Margalit, and A. Yariv. 1982. Low threshold in GaAsP terrace mass transport laser on semi-insulating substrate. *Appl. Phys. Lett.* 41:1115.
    Vawter, G. A., and J. L. Merz. 1984. Design and fabrication of AlGaAs/GaAs phase couplers for OIC applications. *Proceedings of SPIE* 517, p. 15.
    MacDonald, R. I., and D. K. W. Lam. 1985. Optoelectronic switch matrices: recent developments. *Opt. Eng.* 24:220.
28. Hunsperger, R. G. and M. A. Mentzer. 1989. Optical control of microwave devices—A review. *Proceedings of SPIE*, September, Boston, MA.

29. Ury, L., K.Y. Lau, N. Barchaim, and A. Yariv. 1982. Very high frequency GaAlAs laser field effect transistor monolithic integrated circuit. *Appl. Phys. Lett.* 41:126.

30. Matsueda, H. and M. Nakamura. 1984. Monolithic integration of laser diode photomonitor, and electronic circuits on a semi-insumating GaAs substrate. *Appl. Opt.* 23:779.

31. Dakss, M., L. Kuhn, P. F. Keidrich, and B. A. Scott. 1970. *Appl. Phys. Lett.* 16:532.

32. Robertson, M. J., S. Ritchie, and P. Dayan. 1985. Semiconductor waveguides: Analysis of coupling between rib waveguides and optical fibers. *Proceedings of SPIE* 578, p. 184.

33. Hammer, J. M., R. A. Bartolini, A. Miller, and C. C. Neil. 1976. *Appl. Phys. Lett.* 28:192.

34. Bulmer, C. H., and M. G. Wilson. 1976. Single mode grating coupling. *IEEE J. Quant. Electron.* QE-14:741.

35. Cherin, A. H. 1983. *An Introduction to Optical Fibers.* New York: McGraw Hill, pp. 240–241.

36. Hunsperger, R. G., A. Yariv, and A. Lee. 1977. *Appl. Opt.* 16:1–26.

37. Mergerian, D. and E. C. Malarkey. 1980. *Microwave J.* 23:37.
Mergerian, D., E. C. Marlarkey, M. A. Mentzer, et al. 1982. Advanced integrated optic RF spectrum analyzer. *Proceedings of SPIE* 324, p. 149.
Ranganath, T. R. and T. R. Joseph. 1981. Integrated optical circuits for RF spectrum analysis. Final report contract #F33615-78-C-1450. Hughes Research Laboratories for Air Force Wright Aeronautical Laboratories, July.

38. Emkey, W. L. 1983. Optical coupling between single mode semiconductor lasers and strip waveguides. *IEEE J. Lightwave Tech.* LT-1:436.

39. Kingerly, W. D., Bowen, H. K., and Uhlmann, D. R. 1976. *Introduction to Ceramics.* New York: John Wiley & Sons, pp. 685–701.

40. Olsen, G. H., and V.S. Ban. 1987. InGaAs: The next generation of photonic materials. *Solid State Technol.* 30(2):99.
Korotky, S. et al. 1986. Integrated optic narrow line width laser. *Appl. Phys. Lett.* 49(1).

41. Chung, P. S. December 1986. Integrated electro-optic modulators and switches. *J. Electr. Electron. Eng.* Aust.—IE Aust. and IREE Aust. 6(4).

42. Kingerly, Bowen, and Uhlmann, *Introduction to Ceramics.* New York: John Wiley & Sons: 685–701.

43. Boloyaev, R., V. V. Levin, L. E. Marasin, Yu. V. Popov, and L. Yu. Kharberger. 1986. Photorefraction in lithium niobate planar waveguides. *Opt. Spectrosc.* (USSR) 61(1):119–120.

44. Alferness, R. C. 1986. Titanium diffused lithium niobate planar waveguide devices. *ISAF'86 Proceedings,* June 8–11.

45. Burns, W., P. Klein, and E. West. 1986. Ti diffusion in Ti:LiNbO3 optical waveguide fabrication. *J. Appl. Phys.* 60(10).

46. Twigg, M., D. M. Maher, S. Nakahara, T. T. Sheng, and R. J. Holmes. 1987. Study of structural faults in Ti diffused lithium niobate. *Appl. Phys. Lett.* 50(9):501–503.
Bourns, W. et al. 1979. *J. Appl. Phys.* 50:6175.

47. Caruthers, J., I. Kaminow, and L. Stulz. 1974. Diffusion kinetics and optical waveguiding properties of out-diffused layers in $LiNbO_3$. *Appl. Opt.* 13(10):2333–2342.

48. Chung. 1986. Integrated electro-optic modulators and switches. *J. Electrical Electron. Eng.* Bhatt, S., and B. Semwal. 1987. Dielectric and ultrasonic properties of LiNbO3 ceramics. *Solid States Ionics* 23:77–80.

49. Chung, P. S. Integrated electro-optic modulators and switches.

50. Kingerly, W. D., Bowen, H. K., and Uhlmann, D. R. *Introduction to Ceramics.* Alferness, R. C. 1986. Optical guided wave devices. *Science* 234, November 14.

51. Alferness, R. C. Titanium diffused lithium niobate planar waveguide devices.

52. Kingerly, W. D., Bowen, H. K., and Uhlmann, D. R. *Introduction to Ceramics.*

53. Holman, R., L. Altman-Johnson, and D. Skinner. 1986. The desirability of electro-optic ferroelectric materials for guided wave devices. *Proceedings of IEEE* CH2358-0/86/0000-0032.

54. MMT Staff Report. October 1987. Direct write lithography: An overview. *Microelectronics Manufacturing and Testing* 10(11). Sze, S. M. 1985. *Semiconductor Devices Physics and Technology.* New York: John Wiley and Sons.

55. Gray, T. September 1986. Thin film deposition technology for hybrid circuits. *Hybrid Circuit Technology.*

56. Burns, Klein, and West. 1986. Ti diffusion in Ti:LiNbO3 optical waveguide fabrication. *J. Appl. Phys.* Caruthers, Kaminow, and Stulz. Diffusion kinetics and optical waveguiding.

57. Chung. Integrated electro-optic modulators and switches. Armenise, Canalia, and DeSario. 1983. Characterization of $(Ti_{0.65}Li_{0.35})O_2$ compound as a source of Ti during $TiLiNbO_3$ optical waveguide fabrication. *J. Appl. Phys.* 54(1) January. Twigg, M. et al. Study of structural faults in Ti diffused lithium niobate.

58. Burns, W., P. Klein, and E. West. Ti diffusion in Ti:LiNbO3 optical waveguide fabrication. Fontaine, M., A. Delage, and D. Landheer. 1986. Modeling of Ti diffusion into $LiNbO_3$ using a depth dependent diffusion coefficient. *J. Appl. Phys.* 60(7). Palma, F. and L. Schirone. 1986. Acousto-optic interaction efficiency in Ti:LiNbO_3 waveguide collinear Bragg diffraction cell. *J. Appl. Phys.* 60(10).

59. Alferness. 1986. Optical guided wave devices. *Science.* 234(4778): 825–829.

60. Twigg, et al. Study of structural faults in Ti diffused lithium niobate.

61. Armenise, Canalia, and DeSario. 1983. Characterization of $(Ti_{0.65} Li_{0.35})O_2$ compound. *J. Appl. Phys.* 54(1). Fontaine, M., A. Delage, and D. Landheer. Modeling of Ti diffusion into $LiNbO_3$. Hickernell, F., S. Joseph, and K. Ruehle. 1986. Surface wave studies of annealed proton exchanged lithium niobate. *IEEE Proceedings.*

62. Ibid. Gnauck, A. et al. 1986. Personal discussions with author at *Optical Fiber Communication Conference,* February, Atlanta, GA.

63. Burns, Klein, and West. Ti diffusion in Ti:LiNbO3 optical waveguide fabrication.

64. Palma, F. and L. Schirone. 1986. Acousto-optic interaction efficiency. *J. Appl. Phys.* 60(10).

65. Alferness, Titanium diffused lithium niobate waveguide devices. Armenise, M., C. Canalia, and M. DeSario. Characterization of $(Ti_{0.65} Li_{0.35})O_2$ compound. Fontaine, M., A. Delage, and D. Landheer. Modeling of Ti diffusion into $LiNbO_3$.

66. Twigg, et al. Study of structural faults In Ti diffused lithium niobate.

67. Stallard, W. A., A. R. Beaumont, and R. C. Booth. 1986. Integrated optic devices for coherent transmission. *J. Lightwave Tech*. LT-4(7).
    Armenise, M., C. Canalia, and M. DeSario. Characterization of $(Ti_{0.65} Li_{0.35})O_2$ compound.
68. Twigg, et al. Study of structural faults in Ti diffused lithium niobate.
69. Stallard, Beaumont, and Booth. Integrated optic devices.
    Alferness. Titanium diffused lithium niobate waveguide devices.
    Alferness. Optical guided wave devices.
    M. Armenise, C. Canalia, and M. DeSario. 1983. Characterization of TiO_2, $LiNb_3O_8$ and $(Ti_{0.65}Nb_{0.35})O_2$ compound growth observed during TiLiNbO_3 optical waveguide fabrication. *J. Appl. Phys*. 54(11).
70. Twigg, et al. Study of structural faults in Ti diffused lithium niobate.
71. Ibid.
72. Stallard, W. A., A. R. Beaumont, and R. C. Booth. Integrated optic devices.
    Alferness, R. C. Titanium diffused lithium niobate waveguide devices.
    Alferness, R. C. Optical guided wave devices.
73. Ibid.
    Ritchie and Steventon. The potential of semiconductors for optical integrated circuits. British Telecom Research Laboratories, Ipswich, U.K., advance unpublished copy.
74. Eknoyan, O., C. H. Bulmer, R. P. Moeller, W. K. Burns, and K. H. Levin. 1986. Guided wave electro-optic modulators in Ti:LiNbO_3 at 2.6 µm. *J. Appl. Phys*. 59(8):2993–2995.
75. Al-Shukri, et al. 1986. Single mode planar and strip waveguides by proton exchange in lithium tantalate and lithium niobate. *Proceedings of SPIE* 651.
    Glavas, E., P. D. Townsend, G. Droungas, M. Dorey, K. K. Wong, and L. Allen. 1987. Optical damage resistance of ion-implanted LiNbO_3 waveguides. *Electron. Lett*. 23(2).
    Hickernell, F., S. Joseph, and K. Ruehle. Surface wave studies.
76. Wong, K. K. 1985. An experimental study of dilute melt proton exchange waveguides in X- and Z-cult lithium niobate. *GEC J. Res*. 3(4).
    Jackel, J. L., C.E. Rice, and J.J. Veselka. 1982. *Technical Digest (Postdeadline Papers) of Integrated and Guided-Wave Optics*. Pacific Grove, CA, PDP1-1.
    Nutt, A. C. G., K. K. Wong, D. F. Clark, P. J. R. Laybourn, and R. M. De La Rue. 1983. Proton-exchange lithium niobate slab and stripe waveguides: Characterization and comparisons. *Proceedings of Second European Conference on Integrated Optics*, 17–18 October.
    Wong, K. K., A. C. G. Nutt, D. F. Clark, J. Winfield, P. J. R. Laybourn, and R. M. De La Rue. 1986. Characterization of proton-exchange slab optical waveguides in X-cut LiNbO_3. *IEEE Proc. J*. 133(2).
77. Zang, D. and C. Tsai. 1986. Titanium indiffused proton exchanged waveguide lenses in LiNbO_3 for optical image processing. *Appl. Opt*. 25(14).
78. Zang, D. and C. Tsai. 1986. Single mode waveguide microlenses and microlens arrays fabricated in LiNbO_3 for optical information processing. *Appl. Opt*. 25(14).
79. Zang, D. and C. Tsai. Titanium indiffused proton exchanged waveguide lenses.
80. Barnoski, M. K., R. G. Hunsperger, R. G. Wilson, and G. Tangonan. 1975. Proton-implanted GaP optical waveguides. *J. Appl. Phys*. 44:1925–1926.

Mentzer, M. A., R. G. Hunsperger, S. Sriram, et. al. 1985. Ion implanted optical waveguides in gallium arsenide. *Optical Engineering* 24 (2) March/April: 25–29.

Mentzer, M. A., R. G. Hunsperger, S. Sriram, J. M. Zavada, H. A. Jenkinson, and T. J. Gavanis. 1983. Temperature processing effects in proton implanted n-type GaAs. *Appl. Phys.* A32:19–25.

Destefanis, G.L., P. D. Townsend, and J. P. Gailliard. 1978. Optical waveguides in LiNbO$_3$ formed by ion implantation of helium. *Appl. Phys. Lett.* 32:293–294.

Webb, A. P. and P. D. Townsend. 1976. Refractive index profiles induced by ion implantation into silica. *J. Phys.* 9:1343–1354.

Nishimura, T., H. Aritome, K. Masuda, and S. Namba. 1974. Optical waveguides fabricated by B ion implanted fused quartz. Japan. *J. Appl. Phys.* 13:1317–1318.

Valette, S., G. Labrunie, and J. Lizet. 1975. Optical waveguides in ion implanted ZnTe. *J. Appl. Phys.* 46:2731–2732.

Valette, S., G. Labrunie, J. Deutsch, and J. Lizet. 1977. Planar optical waveguides achieved by ion implanted in zinc telluride: general characteristics. *Appl. Opt.* 16:1289–1296.

81. Ziegler, J. F. ed. 1984. *Ion Implantation Science and Technology*. New York: Academic Press, Inc.

82. Poate, J. M. and J. W. Mayer, eds. 1982. *Laser Annealing of Semiconductors*. New York: Academic Press.

83. Mentzer, M. A. 1983. Temperature processing effects in proton implanted n-type GaAs. *Appl. Phys.* A32: 19–25.

Mazey, D. J. and R. S. Nelson. 1969. *Radiation Eff.* 229.

Gamo, K., T. Inada, J. W. Mayer, F. H. Eisen, and C. G. Rhodes. 1987. *Radiation Eff.* 33:85.

Williams, J. S. and H. B. Harrison. 1981. *Laser and Electron Beam Solid Interactions and Material Processing*, J. F. Gibbons, L. D. Hess, and T. W. Sigmon, eds. New York: North-Holland Publishers, p. 209.

84. Williams, J. S. and M. W. Austin. 1980. *Appl. Phys. Lett.* 36:994.

85. Williams, J. S., M. W. Austin, and H. B. Harrison. 1980. *Thin Film Interfaces and Interactions*, J. E. E. Baglin and J. M. Poate, eds. Pennington, NJ: Electrochem. Soc., p. 187.

86. Kular, S. S., B. J. Sealy, K. G. Stephens, D. K. Sadana, and G. R. Booker. 1980. *Solid-State Electron*. 23:831.

87. Eisen, F. H. 1980. *Laser and Electron Beam Processing of Materials*, C. W. White and P. S. Peercey, eds. New York: Academic Press, p. 309.

Donnelly, J. P. 1981. *Nucl. Instrum. Methods* 182/183:553.

88. Kacharin, G. A., N. B. Prichadin, and L. S. Smirnow. 1976. *Sov. Phys.-Semicond.* (Engl. transl.) 9:946.

Golouchenko, J. A. and T. N. C. Venkatesan. 1978. *Appl. Phys. Lett.* 32:464.

Sealy, J., S. S. Kular, K. G. Stephans, R. Croft, and A. Palmer. 1978. *Electron. Lett.* 14:820.

Campisano, S. U., I. Catalanmo, G. Fott, E. Rimini, F. Eisen, and M. A. Nicolet. 1978. *Solid-State Electron*. 21:485.

Barnes, P. A., M. J. Leamy, J. M. Poate, S. D. Ferric, J. S. Williams, and G. K. Celler. 1978. *Appl. Phys. Lett.* 33:965.

89. Williams, J. S. and M. W. Austin. *Appl. Phys. Lett.* 36.

90. Gamo, K., T. Ineda, J. W. Mayer, F. H. Eisen, and C. G. Rhodes. 1977. *Radiat. Eff.* 33:85.

91. Poate, J. M. and J. W. Mayer, eds. 1982. *Laser Annealing of Semiconductors*. New York: Academic Press.
92. Dupois, R. D. 1984. MOCVD of III-V. Semi conductors. *Science* 226:623–629.
93. Kallman, W. R. October 1987. MOCVD epitaxy for MMIC program requirements. *Monolithic Technology*.
94. Dapkus, P. D. 1984. *J. Crystal Growth* 68:345–355.
95. Fraser, A. and A. Rode. October 1987. Making MMICs with a foundry. *Monolithic Technology*.
96. Ibid.
97. Parrish, P. October 1987. Computer-aided engineering in GaAs monolithic technology. *Monolithic Technology*.
98. Damman, C., R. Hulse, and P. Wallace. October 1987. Microwave die sort and wafer mapping for GaAs MMIC manufacturing. *Monolithic Technology*.
99. Ibid.
100. Fraser, A. and A. Rode. Making MMICs with a foundry.
101. Ho, P., T. Andrade, and E. Johnson. August 1987. GaAs MMIC reliability analysis and its impact on microwave systems, part 1. *MSN&CT*.
102. U.S. Government Printing Office. October 1986. *Military Handbook, Reliability Prediction of Electronic Equipment*, MIL-HDBK-217E.
103. Fraser, A. and A. Rode. Making MMICSs with a foundry.
104. Holland, W. R., J. J. Burack, and R. P. Stawicki. 1993. Optical fiber circuits. *Proceedings of 43rd ECTC (IEEE)*, June.
105. Nordin, R. A., W. R. Holland, and M. A. Shahid. 1995. Advanced optical interconnection technology in switching equipment. *J. Lightwave Technol.* 13(6).
106. Grimes, G. J. et al. 1993. Packaging of optoelectronics and passive optics in a high capacity transmission terminal. *Proceedings of 43rd ECTC (IEEE)*, June.
107. Booth, B. Optical Crosslinks, Inc. 2002. Private conversation with author.

6

Optical Diagnostics and Imaging

6.1 Optical Characterization

A variety of specialized test and characterization techniques are employed in the evaluation of optical circuits. Correlation of the information obtained from each characterization provides a complete analysis of the electronic and optical properties of the optical circuit. These are discussed in the following sections.

Figure 6.1 illustrates the experimental setup most commonly used in end-fire coupled loss measurements. Direct end-fire coupling is achieved by focusing laser light into and out of the waveguide with an appropriate lens system. Systems designed for use with solid-state laser diodes eliminate the elliptical beam pattern and astigmation common in these types of lasers and provide micron spot sizes. Mode selective, nondestructive loss measurements can be performed using the prism coupling technique shown in Figure 6.2. Here, light is evanescently coupled from the prism into a waveguide mode at a specific coupling angle determined by the prism geometry and index and the mode order.

Coupling into multiple closely spaced waveguides on a single chip is also often required. This is most easily accomplished by butt-coupling optical fibers to the input waveguide. A chuck is fabricated from a suitable semiconductor material by etching grooves in the material to hold the fibers. Due to precision of the photolithographic techniques used to etch the grooves and precision of standard optical fiber dimensions, this method provides a highly accurate means of aligning the fibers and waveguides.

The output image can be analyzed by a variety of methods. The method illustrated in Figure 6.2 is to sweep the focused image across a detector-slit combination by use of a scanning mirror. The detector output is displayed on an oscilloscope, with its sweep synchronized to the mirror scan. This approach is often modified to remove the sweeping mirror and detector-slit combination by using a precision stepping motor. The detector output and stepping motor position can drive an $X–Y$ plotter to produce a plot similar to the oscilloscope display of the previous setup.

An alternative to these techniques is the use of a two-dimensional (2-D) self-scanning pyroelectric detector array. This device, when coupled with an oscilloscope and computer, permits storage and output of mode profile and

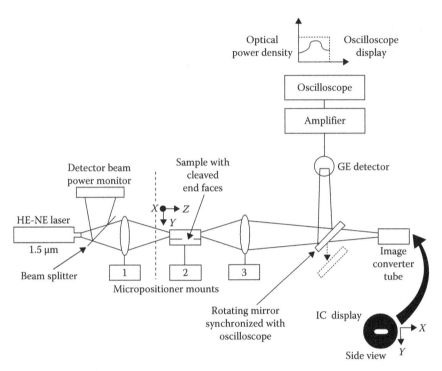

FIGURE 6.1
Mode profile and loss determination setup.

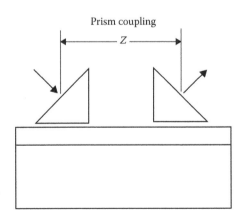

FIGURE 6.2
Prism coupling waveguide loss measurement technique.

loss information. Figure 6.3 illustrates a typical mode profile obtained using such an instrument.

Scattering losses can be measured using a fiber-optic probe technique shown in Figure 6.4. For this technique, micropositioners are accurately controlled to guide a fiber probe along the waveguide (while staying far enough

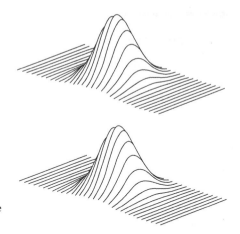

FIGURE 6.3
Representative output from ISC mode profile
and loss determination setup.

away so that coupling is not induced), measuring
any radiation scattered out of the waveguide [1].
The use of the fiber probe incorporated in a microscope allows the distance between the fibers and
the waveguide surface to be kept constant, thereby
improving the accuracy of this technique. The use
of relatively high power AlGaAs lasers, which are
now commercially available, makes this technique
quite viable, as more power input results in more
scattered power.

Several other techniques exist for measuring waveguide propagation loss. A generaliza-
tion of the fiber probe measurement is to use a
video camera to image scattered radiation along
the propagation path; then, by employing digital
image processing, the propagation loss can be
calculated [2]. To use this technique with AlGaAs
waveguides, an infrared video camera is often

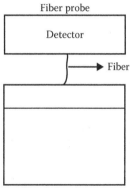

Fiber probe

Detector

→ Fiber

– Out of plane scattered
light proportional to loss

FIGURE 6.4
Fiber probe loss measure-
ment technique.

used. Another possibility to measure waveguide propagation loss is the
use of Fabry–Perot techniques [3]. Here the optical length of the waveguide or the wavelength of the source is varied over a small range with
the output intensity being monitored. The optical length can be changed
by slightly heating or cooling the sample, while varying the wavelength
of the laser source over a small range by inducing the appropriate variations in the power supply current. Radiation losses are measured in bends
in a manner similar to the absorptions loss; however, by calibrating and
subtracting the other two loss mechanisms, the loss due to radiation alone
can be evaluated.

6.2 Bandwidth Measurement

A key characteristic of optical switches and modulators is their switching speed and modulation depth. For switches, the bandwidth is related to the switching speed by the equation

$$T = \frac{2\pi}{\Delta f} \tag{6.1}$$

A modulator's bandwidth is often expressed as the frequency range over which it can maintain a modulation depth of 50%. Various expressions are used to express the modulation depth, depending on whether the applied electrical signal acts to increase or decrease the intensity of the transmitted light. The respective expressions are given as Equation 6.2:

$$n = \frac{(l_0 - l)}{l_0} \text{ or } n = \frac{(l - l_0)}{l_m} \tag{6.2}$$

where
l is the transmitted optical intensity
l_0 is the quiescent optical intensity
l_m is the optical intensity under maximum electrical signal

To test the active devices for modulation depth and speed, high-speed detectors and storage oscilloscopes are used.

6.3 Stability: Temperature and Time Effects

The most significant problem of the stability of waveguide building blocks and devices is introduced by the temperature dependence of the index of refraction resulting from change in the energy bandgap with temperature. Propagation and attenuation of optical energy in a semiconductor can be described by a complex index of refraction

$$n_c = n - ik \tag{6.3}$$

where the real part of the index is related to the propagation constant β by

$$n = \frac{2\pi\beta}{\lambda_0} \tag{6.4}$$

and k is related to the absorption coefficient α by

$$\alpha = \frac{4\pi v k}{c} \tag{6.5}$$

where
   v is the light frequency
   c is the speed of light in a vacuum
   λ_0 is the vacuum wavelength of the light

The real and imaginary parts of the complex index of refraction are coupled to each other as described by the Kramer–Kronig relations [4]. This coupling results in a change in the real part of the index n corresponding to any change in the imaginary part k. At wavelengths relatively close to the absorption edge of the semiconductor (i.e., photon energies close to the bandgap energy E_g), the strong dependence of the optical absorption of E_g produces a corresponding dependence of n on E_g. Thus, the dependence of E_g on temperature translates into a dependence of n on temperature. It is possible to obtain an order-of-magnitude estimate of the effect by using some well-known empirical relations. An empirical relation

$$n^4 E_g \ (\text{eV}) = 77 \tag{6.6}$$

which is called Moss's rule has been found to be obeyed by semiconductors whose value of n^4 is in the range of 30–440 [5]. Note that for $Al_x Ga_{(1-x)} As$, with $x=0$ to $x=0.4$, n^4 is in the range of 164–120. An empirical relation for the bandgap energy of GaAs is given by [6]

$$E_g \ (\text{eV}) = 1.51 - 2.67 \times 10^{-4} T \ (\text{K}) \tag{6.7}$$

for temperatures above 100 K. Using these relations, one would expect an increase in index of refraction of $\Delta n \sim 0.01$ (about 0.3%) for an increase in temperature from 273 to 350 K. This is a relatively small change in index compared to the difference in index between the waveguide and confining layers of an AlGaAs waveguide; hence, little change in the light propagation characteristics would be expected. Also, the decrease in bandgap energy from 1.437 to 1.417 eV as the temperature is increased to 350 K should not produce a significant increase in interband absorption, and free-carrier concentration (and hence absorption) would not change significantly. The change in index, however, even though relatively small, could be expected to have a significant effect on interference-based devices such as a Mach–Zehnder interferometer, and operating temperature would have to be taken into account in their design. Individual devices should be tested in situ at various temperatures to determine if temperature variations are a factor affecting the performance of the device.

The long-term effects of time and temperature on waveguide and device operation are difficult to predict theoretically. It is usually necessary to experimentally determine such performance data by conducting life tests on the various devices after they are fabricated.

6.4 Measurement of $N_D(d)$ Using Capacitance–Voltage Technique

The capacitance–voltage (C-V) technique is an effective method for determining the dopant profile as well as the width of carrier compensated guiding layers, Schottky barriers, p–n junctions, and heterojunctions. The thickness of the depletion layer is determined from the capacitance of a non-injecting metal contact on the guide surface. The voltage V_p required to form a depletion layer of thickness d in a conducting layer of carrier concentration N_D and ϵ_r is given by

$$V_p = \frac{qN_D d^2}{2\epsilon_r} \tag{6.8}$$

If the bulk breakdown field E_c of the semiconductor is exceeded during the measurement, current flows, the accuracy of the technique is immediately seriously impaired, and the measurement rapidly fails. This occurs when

$$E_c = \frac{qN_D d}{\epsilon_r} \tag{6.9}$$

and limits d to the value

$$d = \frac{\epsilon_r E_c}{qN_D} \tag{6.10}$$

where E_c is approximately 6×10^5 V/cm for $N_D = 10^{17}$. This limits the measurable layer thickness to the values indicated in Figure 6.5. Chemical etching sequences are required to characterize deeper layers. Figure 6.6 shows an automated CV measurement system.

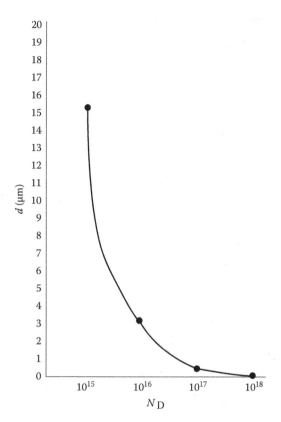

FIGURE 6.5
Thickness of proton implanted waveguide versus carrier concentration, which can be evaluated by a direct CV profile.

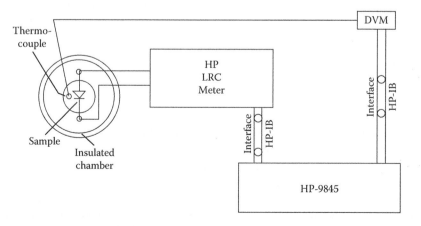

FIGURE 6.6
Block diagram of C-V measurement system.

6.5 "Post Office" Profiling

This system overcomes some of the limitations of the CV technique. The system was originally developed at the British Post Office Research Centre by Gyles Webster [7] to profile carrier concentration in III–V materials. The semiconductor sample is loaded onto a PTFE electrochemical cell where a small area (typically 0.1 or 0.01 cm²) is defined and used to form an electrode. A DC voltage is applied to produce a bias which causes the formation of a Schottky barrier at the material surface; simultaneously, continuous electro-etching takes place under illumination. Capacitance and voltage are periodically monitored, in situ, as the etch progresses, in automatic fashion. A continuous profile is obtained, and a plot of log (carrier concentration versus depth) is drawn.

6.6 Spreading Resistance Profiling

The two-point spreading resistance measurement can be used to map the resistivity of a structure as a function of depth. This can provide information concerning the degree of implant damage produced as well as uniformity and abruptness of transition regions. This is in addition to the standard four-point probe measurement of resistivity shown in Figure 6.7.

The spreading resistivity measurement is made with two probes contacting a surface of the material. At least one of the probes is so sharp as to contact the surface with a small circular area of radius r. Applying a voltage across the electrodes causes a flow of current, whose value is restricted by the resistance of the sharp contact. (Often, both contacts are sharp.) This resistance, called the spreading resistance, is caused by the concentration of the current density in the vicinity of the sharp contact and is thus localized at that point of measurement. Its value is given by

$$R = \frac{\rho}{2\pi r} \tag{6.11}$$

where ρ is the resistivity (Ω-cm) in the neighborhood of the probe.

Figure 6.8 shows schematically how the spreading resistance measurement serves to map the carrier concentration in implanted samples. After exposure the samples are cut into test specimens which are cleaned and then lapped to provide an angle of about 5.7° (tangent = 0.1). This angle provides a 10:1 distance magnification in the measurement. The measurements start with both electrodes positioned on the line where the lapped surface meets the original surface at an angle of about 174.3°. After providing a value of the

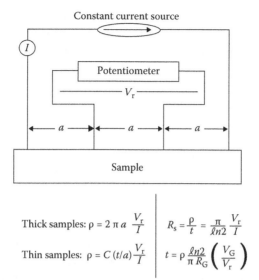

FIGURE 6.7
Four-point probe measurement of resistivity.

Thick samples: $\rho = 2\pi a \dfrac{V_r}{I}$ $\qquad R_s = \dfrac{\rho}{t} = \dfrac{\pi}{\ell n 2}\dfrac{V_r}{I}$

Thin samples: $\rho = C\,(t/a)\dfrac{V_r}{I}$ $\qquad t = \rho\,\dfrac{\ell n 2}{\pi R_G}\left(\dfrac{V_G}{V_r}\right)$

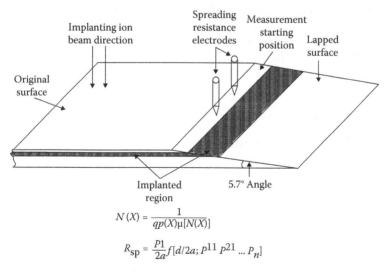

$$N(X) = \frac{1}{qp(X)\mu[N(X)]}$$

$$R_{sp} = \frac{P1}{2a}f[d/2a;\, p^{11}\,p^{21}\,...\,P_n]$$

FIGURE 6.8
Spreading resistance measurement.

resistance R at that location, the electrodes are advanced to the right in small steps. Steps of 5 µm, for example, correspond to measurements at depth increments of 0.5 µm.

The expected resistivities for GaAs tests are much less than for silicon. They run from about 0.12×10^{-2} to 8.8×10^{-2} Ω-cm in undamaged and damaged regions, respectively. These values are calculated using values of electron mobility from Blakemore [8] and corresponding anticipated free carrier concentrations of between 10^{16} and 10^{18} cm^{-3}. Under the assumption that the effective radius of the probe is $0.75\,\mu m$, the corresponding spreading resistance values range between 4 and $186\,\Omega$. The following table illustrates these relations.

Degree of Damage	None	Some	Full
Electron concentration (cm^{-3})	10^{18}	10^{17}	10^{16}
Mobility (cm^2/V/s)	3300	5300	7100
Resistivity (Ω-cm)	0.19×10^{-2}	1.2×10^{-2}	8.8×10^{-2}
Spreading resistance (Ω)	4	25	186

There are several approaches which can be used in applying spreading resistance measurements to study damage in GaAs optical channel waveguides, for example. The easiest and most straightforward is to expose a separate unmasked sample of GaAs to the incident beam simultaneously with the exposure of the main pattern sample. The separate sample can then be sacrificed and studied separately. Alternatively, the separate sample could be a selected region on the main wafer, to be split off and studied separately. It is also possible, though more tedious, to sacrifice and study the actual channels with this method. The major difficulty is in achieving adequate alignment between the channel, the lapping edges, and the electrodes.

6.7 Mobility Measurement

Conwell and Weisskopf [9] determined a relationship between the carrier concentration in a semiconductor and the carrier mobility:

$$\mu = \frac{64\sqrt{\pi}\epsilon_s^2 (2kT)^{\frac{3}{2}}}{Nq^3 (m^*)^{\frac{1}{2}}} \left\{ \ln\left[1 + \left(\frac{12\pi\epsilon_s kT}{q^2 N^{\frac{1}{3}}}\right)^2\right]\right\}^{-1} \tag{6.12}$$

A differential four-point van der Pauw measurement (Figure 6.9) can be performed so that successive chemical etches of the wafer (each of known thickness) will provide $u(d)$. The $N_D(d)$ information obtained with a Post Office profile provides

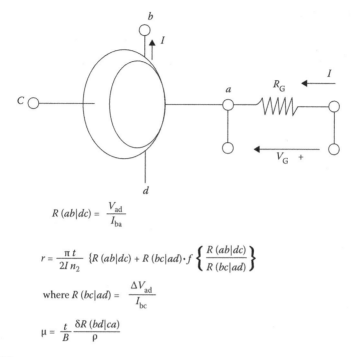

$$R\,(ab|dc) = \frac{V_{\text{ad}}}{I_{\text{ba}}}$$

$$r = \frac{\pi\,t}{2I\,n_2}\,\{R\,(ab|dc) + R\,(bc|ad)\}\cdot f\left\{\frac{R\,(ab|dc)}{R\,(bc|ad)}\right\}$$

$$\text{where } R\,(bc|ad) = \frac{\Delta V_{\text{ad}}}{I_{\text{bc}}}$$

$$\mu = \frac{t}{B}\,\frac{\delta R\,(bd|ca)}{\rho}$$

FIGURE 6.9
Van der Pauw measurement of resistivity and carrier mobility.

$$n_1(x) - n_2 = \frac{e^2}{2n_2\epsilon_0 m^*\omega^2}\big[N_D - N_1(x)\big] \tag{6.13}$$

We then know both $N_D(d)$ and $u(d)$, from which we can derive N_D versus u.

An attractive experiment for measurement of the drift mobility of the carriers in implanted layers is the "time of flight" measurement. In this technique [10], "a relatively compact group of excess carriers, released or injected by some form of impulsive excitation, is caused by an externally applied electric field to traverse a known distance in more or less coherent fashion, and the time of its arrival is measured. In the absence of all diffusion, recombination, relaxation, and deep trapping effects, the drift mobility is given by $\mu = d/E\tau$, where d is the known distance, E is the applied field, and τ is the observed transit time." Techniques are currently under investigation to transform the experiment from the time domain to the frequency domain. It is believed that the "Fourier transform" method of drift mobility measurement will have practical advantages over the time-of-flight method.

6.8 Cross-Section Transmission Electron Microscopy

Microscopy is a very sensitive observational technique used to routinely investigate surface morphology and cleaved sections of multilayer structures. In conjunction with potassium hydroxide treatments, it allows the identification of large antiphase domains and stacking faults and the measurement of dislocation density.

This technique is often used to obtain the carrier distribution in implanted guides. It can potentially provide both transverse and longitudinal carrier distributions. Grinding and polishing are performed using a resin/semiconductor composite to reduce the sample thickness to 4–6 mils. Samples are then ion milled to ~2000 Å for TEM analysis and stacked so that the implanted layers are perpendicular to the plane of the sample.

In order to experimentally determine the location of implanted damage in GaAs, specimens are annealed prior to STEM. The annealing step allows defect migration and formation of extended line and loop defects which can be imaged by STEM. The post-anneal damage profile can then be delineated by calculating defect densities as a function of depth. Although the width of the peak damage region will be reduced somewhat by the anneal, the peak damage and end of range values will essentially remain unchanged.

6.9 Infrared Reflectivity Measurements

Infrared reflectivity of waveguides is often performed using a specular reflective attachment to a spectrophotometer [11]. The thickness of the waveguide is, to a good approximation, given by

$$t = \frac{1}{2s(n_f^2 - \sin^2\theta)^{\frac{1}{2}}} \tag{6.14}$$

where
  s denotes the average spectral separation (in wave numbers) between maxima and minima in the fringe pattern
  θ is the angle of incidence of the light rays
  n_f is the average value of the refractive index of the film (see Figure 6.10)

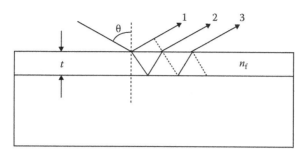

FIGURE 6.10
Interference measurement of dielectric film thickness.

6.10 Other Analysis Techniques

Angle polishing and staining is a technique enabling the observation of material layers down to a thickness on the order of 100 Å (see Figure 6.11). Surface profiling is accomplished with various profilometers. These devices allow routine assessment of surface roughness to a resolution of 0.1 µm. The electron microprobe is a device used to determine alloy composition of AlGaAs materials in the indirect band gap regime (i.e., for percentages of aluminum greater than approximately 45%). Hall and photo-Hall measurements yield information on material carrier density and mobility.

Deep-level transient spectroscopy (DLTS) and capacitance–transient (C-T) measurements are commonly used to provide a wide range of information

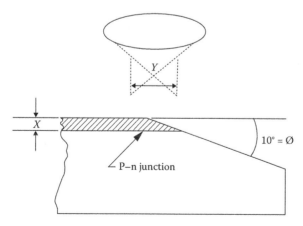

FIGURE 6.11
Measurement of junction depth by angle polishing and staining.

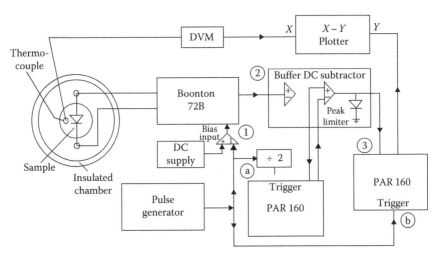

FIGURE 6.12
Block diagram of the DLTS measurement system.

on the electronic and optical defects in a material. These two capacitance techniques provide information on defect activation energy, capture cross section, and density. Figures 6.12 and 6.13 show the DLTS and C-T block diagrams.

Photoluminescence is an invaluable tool providing information on crystal uniformity, impurities, optical efficiency, and energy band gap for direct-gap

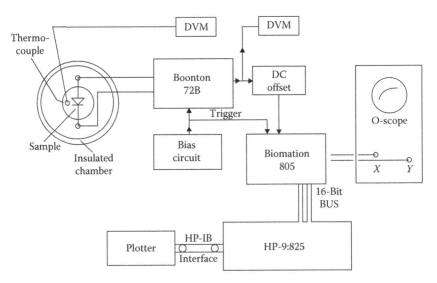

FIGURE 6.13
Block diagram of the C-T measurement system.

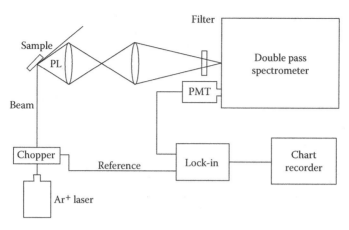

FIGURE 6.14
Photoluminescence system.

semiconductors. The analysis of the half-width and intensity of AlGaAs system quantum well structures provides a very sensitive indication of the microscopic interface roughness of III–V heterostructures. Figure 6.14 shows the photoluminescence system block diagram.

6.11 Biotechnology Applications

Na, K, and Cl are key ions inside and outside the cell, providing for fundamental cell function. Proteins with pump or channel features is now an accepted description for the mediating mechanism that regulates ion partitioning cellular function [12]. Receptor proteins may be closely linked to pump and channel proteins and may act to modulate their activity. Methods such as x-ray diffraction (see Chapter 7) provide valuable insights into protein structure and function. These are complemented by biological analysis tools such as sodium dodecylsulfate polyacrylamide gel electrophoresis (SDS-PAGE).

6.12 Parametric Analysis of Video

Video data analysis requires proper camera setup, timing, triggering, calibration, data acquisition, storage and transfer, and ultimate parametric analysis and reporting. This is accomplished in a seamless workflow with optimal efficiencies, ensuring digital image data is optimized for parametric analysis and event characterization. Ancillary technologies include advanced illumination techniques, including high brightness laser illumination, arc discharge lamp illumination, plasma lighting, structured lighting, and signal enhancement techniques. Custom MATLAB® routines are often employed for analysis of x-ray imagery. Specialized image acquisition methodologies are enhanced with data fusion from fiber-optic sensors and strain gauges, providing enhanced event characterization.

A typical process flow is illustrated in Figure 6.15 for an event employing, for example, high-speed video acquisition, x-ray cineradiography, surface deformation, and an integrated sensor suite. Event characterization may include analysis of impact velocity and velocity curves, pitch and yaw during trajectory and at impact, and deformation, providing correlation to a range of parameters.

A typical problem uses an image as an intermediate data structure, formed from raw data, which is analyzed further to extract desired information. This information—such as the location of a defect in a body armor image—can best be determined only after thoroughly understanding the physics of the imaging process and characterizing all available prior knowledge. Given this understanding, we develop signal processing algorithms that involve concepts from random signals, statistical inference, optimization theory, and digital signal processing [13].

A host of high-speed cameras are available commercially, including Vision Research Phantoms and Photron high-speed cameras (see Figure 6.16).

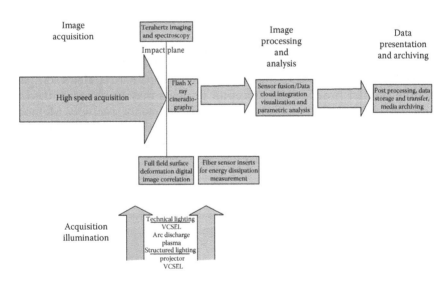

FIGURE 6.15
High-speed video acquisition process.

(a)   (b)

FIGURE 6.16
Representative high-speed video cameras. (a) Photron SA-5 (b) Vision Research V1.3.

6.12.1 Parametric Analysis of Digital Imagery

Digital image processing suites available for commercial licensing include ARAMIS, TrackEye, ProAnalyst, Correlated Solutions, PixelRay, and developmental MATLAB routines. Customized routines for specific test applications are available, along with data fusion processes for enhanced visualization and accuracy.

Two-dimensional parametric analysis requires a single imager with a calibration sample in the plane of motion either before or during the event. This provides measurement of 2-D flight characteristics such as pitch, yaw, position, velocity, and acceleration as well as angle, distance, and area, using the interactive feature tracking overlays. Distance and velocity, for example, gun bolt movement or the height an object moves into the air from a mine blast, are readily determined from 2-D imagery.

Three-dimensional (3-D) parametric analysis requires at least two imagers and a 3-D calibration fixture. This provides 3-D flight characteristics such as pitch, yaw, position, velocity, and acceleration as well as angle, distance, and area. Coordinate location and distance calculation for a traveling object are derived from the world reference points collected in a 3-D calibration. This is compared with time or fused with data from additional sensors.

Pitch and yaw for slugs and projectiles traversing a known path are calculated with a single 2-D image or, with 2 cameras, a 3-D image of the projectile flight.

Area and volume for smoke cloud or flash evaluation are performed in 2-D and utilize a 2-D calibration in the event plane. Volume measurements utilize 3-D analysis and calibration.

Surface deformation/3-D image correlation can be used for measurement of backface deformation during armor tests (exploitation). Surface deformation analysis utilizes a uniform random black and white speckle pattern on the surface and two calibrated high-speed cameras. Full field deformation analysis provides displacement and velocity in three coordinates, x–y strain, principal strain angle, in-plane rotation, 3-D contours, and curvature angles.

FIGURE 6.17
ARAMIS System. (Courtesy of Gesellschaft Für Optische Messtechnik [GOM] mbH, Braunschweig, Germany.)

Complementary solutions in development include projected structured lighting, laser illumination/bandpass filtering, and specialized illumination techniques.

An excellent example of optical 3-D measurement of surface deformation is the ARAMIS system produced by the Gesellschaft für Optische Messtechnik, GOM mbh (see Figure 6.17). Following is an excerpt from the GOM literature, illustrating the range of optical measurement capability for the 3-D software suite.

ARAMIS helps to better understand material and component behavior and is ideally suited to monitor experiments with high temporal and local resolution.

ARAMIS is a noncontact and material-independent measuring system providing, for static or dynamically loaded test objects, accurate

- 3-D surface coordinates
- 3-D displacements and velocities
- Surface strain values (major and minor strain, thickness reduction)
- Strain rates

ARAMIS is the ideal solution for

- Determination of material properties (R- and N-values, FLC, Young's Modulus, etc.)
- Component analysis (crash tests, vibration analysis, durability studies, etc.)
- Verification of finite element analysis

ARAMIS is the unique solution delivering complete 3-D surface, displacement and strain results where a large number of traditional measuring devices are required (strain gauges, LVDTs extensometers, etc.).

The same system setup is used for multiple applications and can be easily integrated in existing testing environments (see Figure 6.19).

6.12.1.1 CAD Data Integration

ARAMIS provides an import interface for CAD data, which are used for 3-D coordinate transformations and 3-D shape deviation calculations. (See Figure 6.18.)

The import interface handles the following formats:

- Native: Catia v4/v5, UG, ProE
- General: IGES, STL, VDA, STEP

6.12.1.2 Real-Time Data Processing

The ARAMIS software provides real-time results for multiple measurement positions from the test object's surface.

These are directly transferred to testing devices, data acquisition units, or processing softwares (e.g., LabView, DIAdem, and MSExcel) and are used for

- Controlling of testing devices
- Long-term tests with smallest storage requirements
- Vibration analysis
- 3-D video extensometer

6.12.1.3 Verification of FE Simulations

As part of complex process chains, optical measuring systems have become important tools in industrial processes in the last years. Together with the numerical simulation they have significant potential for quality improvement and optimization of development time for products and production.

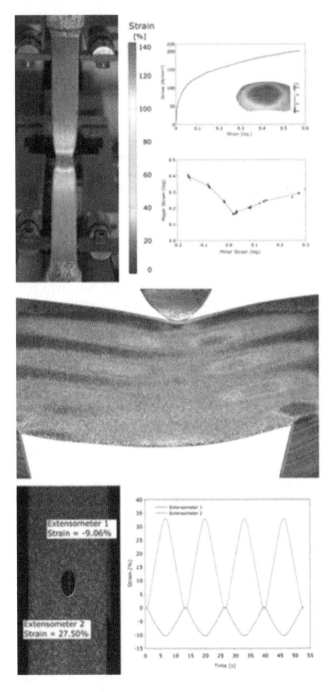

FIGURE 6.18
Representative parametric analysis visualizations. (Courtesy of Gesellschaft Für Optische Messtechnik [GOM] mbH, Braunschweig, Germany.)

ARAMIS strongly supports the full-field verification of FE-simulations. Determining material parameters with ARAMIS helps to evaluate and improve existing material models. The import of FE result datasets allows one to perform numerical full-field comparisons to FE simulations for all kinds of component tests. Thus finite element simulations can be optimized and are getting more reliable).

6.12.1.4 Complete Workflow in One Software Application

The entire measuring, evaluation and documentation process is carried out within the integrated ARAMIS software.

The full potential of the available hardware is used to capture and evaluate the measuring area efficiently and with high accuracy (see Figure 6.19).

6.13 X-Ray Imaging

X-ray imaging applications require unique optical detector system configurations for optimization of image quality, resolution, and contrast ratio. X-ray photons from multiple anode sources create a series of repetitive images on fast-decay scintillator screens, from which an intensified image is collected as a cineradiographic video (or intensified image) with high-speed videography Current developments address scintillator material formulation, flash x-ray implementation, image intensification, and high-speed video processing and display. Determination of optimal scintillator absorption, x-ray energy and dose relationships, contrast ratio determination, and test result interpretation are necessary for system optimization [14].

6.13.1 Introduction

Numerous test requirements motivate the development of flash x-ray cineradiography systems with multi-anode configuration for repetitive imaging in closely spaced timeframes. Applications include the following:

- Shaped charge detonations to further understand properties of jet formation and particulation
- Explosively formed projectile detonations to quantify launch and flight performance characteristics
- Detonations of small caliber grenades and explosive projectiles to verify fuse function times and fragmentation patterns
- Performance and behavior of various projectiles and explosive threats against passive, reactive, and active target systems
- Human effects studies including body armor, helmets, and footwear

Technical data				
System configurations	5M	4M	HS	High Speed
Frame rate (Hz)	up to 15 (29)	up to 60 (480)	up to 500 (4,000)	up to 1,000,000
Camera resolution (pixel)	2,448 × 2,050	2,358 × 1,728	1,280 × 1,024	up to 1,024 × 1,024
Measuring area	mm^2 to $>m^2$	mm^2 to $>m^2$	mm^2 to $>m^2$	mm^2 to $>m^2$
Strain measuring range (%)	0.01 up to >100	0.01 up to >100	0.01 up to >100	0.01 up to >100
Strain measuring accuracy (%)	up to 0.01	up to 0.01	up to 0.01	up to 0.01
Camera base	variable/ fixed	variable/ fixed	variable/ fixed	variable
Tool free mounting	▪	▪	▪	–
Integrated cable Guide	▪	▪	▪	–
Positioning pointers	1 or 3	1 or 3	1 or 3	–
Illumination	integrated	integrated	external	external
High-end 19" PC	▪	▪	▪	▪
Note book	▪	–	–	▪
Control device	sensor controller	sensor controller	sensor controller	optional
Sensor dimensions (high × depth in mm)	175 × 180	235 × 175	230 × 185	
Sensor dimensions (length)	500, 800, ...	500, 800, ...	500, 800, ...	
Specimen temperature		typ. −100°C	up to +1,500°C	
Weight (kg)	5	6.5	6	

FIGURE 6.19
ARAMIS analysis system examples. (Courtesy of Gesellschaft Für Optische Messtechnik [GOM] mbH, Braunschweig, Germany.)

- Behind armor debris studies of large caliber ammunition against various armor materials
- Small caliber projectile firings to study launch, free flight, and target impact results

Specific parameters of interest include impact velocity of projectiles striking various materials, pitch, yaw, and roll during projectile trajectory and at impact, time to maximum deformation of materials, time to final relaxation state, and profile of the deformation occurring during the events of interest.

6.13.2 Flash X-Ray

Flash x-ray is a well-established technology in which an electron beam diode (anode and cathode) in a sealed tube produces a brief (<30 ns) intense x-ray pulse. Recovery times are too slow to repetitively pulse the anode for high-speed cineradiography—high frame speed requires multiple independent pulsers and diodes. Sources are closely packed to minimize parallax problems and to maintain source intensity with smaller diodes [15]. Triggering of x-ray pulses is tailored to the event timing, so that precise imaging of the event is achievable with either high-speed framing cameras in which image frames are matched to the x-ray pulses, or high-speed video cameras with suitable exposure rates. Image processing algorithms are then utilized to allow extraction of parametric data from multiple frame x-ray images produced on scintillator screens and images through microchannel plate or image intensifier hybrids.

X-ray computed tomography (XCT) represents an alternative approach with the advantage of a 3-D image. The technique requires four or more pulsers for each 3-D frame, translating to 32 or more pulsers, to produce an equivalent eight frames [16]. There is no commercially available anode arrangement that lends itself to the correct geometry, and the cost of such a system would be substantial. For this reason, custom systems are constructed for specific applications.

6.13.3 Typical System Requirements

System specifications for cineradiography systems in development include the following:

- Multi-anode x-ray system, 150–450 keV x-ray photons
- Fluorescent screen/real-time video camera imaging, 6–8 foot standoff
- One to five mm spot size, with $1\,m^2$ target area
- Up to 4 ms record time
- Eight images at 100,000 frames per second, trigger synch pulses
- Pulser repetition rate adjustable from 10,000 to $100,000\,s^{-1}$
- Independent inter-pulse time from anode to anode
- Adaptable to a wide range of test scenarios/operation in harsh conditions/explosive test environment

Shown in Figure 6.20 is a multi-anode arrangement where the acquisition is achieved with a fast scintillator and a high-speed video camera. Multiple frames of imagery are produced on a scintillator screen by flash x-ray exposure through the test materials at either 150 or 450 keV. A turning mirror is used to avoid placement of the video camera directly in the residual x-ray beam. In a former approach, separate x-ray heads were

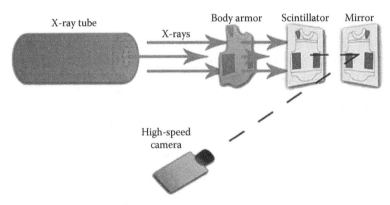

FIGURE 6.20
Typical multi-anode system configuration.

used to image the event from different directions without overlap of the projection onto storage phosphor screens or x-ray film, from which still video images were obtained. This approach creates parallax and geometry issues, thus motivating the current approach in which successive frames are captured on a single area location on a scintillator screen. A framing camera with an image intensifier is synchronized with the x-ray pulses to capture snapshots over events of duration from tens of microseconds to a few milliseconds.

The choice between 150 and 450 keV x-rays represents a tradeoff between material penetration and contrast ratio. The attenuation coefficient for clay, for example, as reported by NIST [17] and Aberdeen, is 0.2 cm^{-1} at 450 keV and 0.3 cm^{-1} at 150 keV. For 10 in. of clay, the 150 keV flux is attenuated by about 15X with respect to a 450 keV pulse. It is possible, using 450 keV x-rays and a storage phosphor screen, to image a 0.5 in. depression under the described angular alignment. The attenuation differential with x-ray energy implies a contrast ratio difference. As it is desirable to image depressions smaller than 0.5 in. (1.27 cm), with optimal resolution, the improved contrast ratio that obtains for 150 keV x-rays is an advantage—if the higher flux requirement can be accomodated.

6.13.4 Scintillators for Flash X-Ray

Six commercially available fast scintillators, from two manufacturers and one distributor, were characterized for efficiency and speed at 150 and 450 keV, as efficiency differences impact the increased flux requirements for imaging at 150 keV. No information on uniformity or sensitivity was available, and the speed of the samples deviated from data published in the literature. Availability, pricing, and timely delivery of suitable scintillators is expected to be a challenge associated with assembly of multiple systems.

The scintillators had various sizes up to about 6 in., so they were masked to have a common 6 mm diameter circular area from which light could be collected. Each was placed in a dark chamber with a fast photomultiplier tube module (Hamamatsu H9656MOD), containing a transimpedance amplifier. X-rays from the pulsers were directed at the scintillators from about 8 in. outside of the dark box, at a flux density of the order of 10^9 cm^{-2} and duration of approximately 30 ns/pulse. The photomultiplier was located about fourteen inches from the scintillator, and was followed by a Tektronix TDS 2024B 200 MHz 2 Gs/s oscilloscope. Photographs of the resulting traces were used to report the integrated output and speed of the scintillators.

It is interesting to note that the efficiency at 150 keV is not very different from that at 450 keV, despite the 7.4× higher attenuation reported in the literature for 150 keV with respect to 511 keV [17]. Future work around the visible photon yield per x-ray, as a function of x-ray energy and scintillator flavor, may help clarify the issue. It is also notable that the measured decay times are far longer than those reported [18]. For frames of about 1 µs duration, placed at various spacing throughout an event of tens to thousands of microseconds, the scintillator response times are inadequate.

6.13.5 System Configurations

Using currently available technology, the apparatus typically required in order to implement a flash cinematography system will likely include

- High-voltage power supply
- Trigger, timing, and delay control electronics
- Trigger amplifier
- X-ray pulser(s) + [1] multi-anode x-ray tube
- Target positioning and backing fixture
- High decay rate scintillating screen (<10 µs decay)
- Mirror
- Image intensifier
- High frame rate camera
- Computer control system and software
- Connecting cables

This equipment can be configured in order to provide an efficient and safe means of capturing x-ray image sequences synchronized to high-velocity test events. The use of a multi-anode tube for x-ray exposure of fixed position targets will reduce the complexity of some test arrangements. Complete characterization of any parallax effects from multiple anode spacing can be used to enhance spatial resolution of images used for metrology. The imaging system selected for such a test configuration will require careful consideration of scintillating screen fluorescence, decay characteristics, and resolution limits. Similarly, the image intensifier resolution characteristics should be well matched to the system application. The camera utilized for image capture of the test sequence may be subject to upgrades as technical advances permit higher sensitivity and reduced exposure time.

Further development of flash x-ray cinematography includes the possibility of using directly illuminated CCD based detectors, UV scintillating screens, and x-ray diffraction apertures to control and/or reduce x-ray exposure regions [19,20]. Further advances in spatial metrology methods (such as point cloud algorithms) will further enhance the analytical capabilities for flash x-ray cinematography.

References

1. Developed by Boyd, J. University of Cincinnati. Private conversation with author.
2. Okamura, Y., S. Yoshinaka, and S. Yamamoto. 1983. Measuring mode propagation losses of integrated optical waveguides: A simple method. *Appl. Opt.* 22:3892.
3. Walker, R. G. 1985. Simple and accurate loss measurement techniques of semiconductor optical waveguides. *Electron. Lett.* 21:57.
4. See for example, Pankove, J. I. 1971. *Optical Properties of Semiconductors.* Englewood Cliffs, NJ: Prentice Hall, pp. 89–90.
5. Moss, T. S. 1959. *Optical Properties of Semiconductors.* New York: Butterworth, pp. 59–88.
6. Gooch, C. H. 1959. *Gallium Arsenide Lasers.* New York: Wiley-Interscience, p. 48.
7. Webster, G. 1983. Polaron Corporation. Private communication with author.
8. Blakemore, J. S. 1982. Semiconductor and other major properties of galluim arsenide. *J. Appl. Phys.* 53 (10):123.
9. Conwell, E. and V. F. Weisskopf. 1950. Theory of impurity scattering in semiconductors. *Phys. Rev.* 77:388.
10. Fagen, E. A. University of Delaware. 1984. Private conversation with author. Haynes, J. R. and W. Schockley. March 1951. The mobility and life of injected holes and electrons in germanium. *Phys. Rev.* 81:835–843.
11. Zavada, J. M., H. A. Jenkinson, T. J. Gavanis, R. G. Hunsperger, M. A. Mentzer, D. C. Larson, and J. Comas. 1980. Temperature processing effects in proton-implanted GaAs. *Proceedings of SPIE* 239. p. 157F.
12. Pollack, G. 2001. *Cells, Gels and the Engines of Life.* Seattle: Ebner and Sons Publishers.
13. Prince, J. L. and M. L. Jonathan. 2006. *Medical Imaging, Signals and Systems.* Upper Saddle River, NJ: Pearson Prentice Hall.
14. Mentzer, M. A., D. A. Herr, K. J. Brewer, N. Ojason, and H. A. Tarpine. January 2010. Detector development for x-ray imaging. *Paper Presented at SPIE Photonics West*, Sanfrancisco, CA.

15. Mattsson, A. 2007. New developments in flash radiography. *Proceedings of SPIE 27th International Congress on High-Speed Photography and Photonics*, China, 6279, 62790Z.

16. Karsten, M. and P. Helberg. 2005. Computed tomography of high-speed events. *Proceedings of 22nd International Symposium on Ballistics* 2, Vancouver, BC, Canada.

17. National Institute of Standards and Technology Physics Laboratory. 2010. Photon cross sections, attenuation coefficients, and energy absorptions coefficients from 10 keV to 100 GeV. NSRDS-NBS 29. U. S. Commerce Department. http://physics.nist.gov/PhysRefData/XrayMassCoef/ElemTab/z06.html

18. Shionoya, S. and W. M. Yen, eds. 1999. *Phosphor Handbook*. Boca Raton, FL: CRC Press.

19. Helberg, P., S. Nau, and K. Thoma. 2006. High-speed flash x-ray cinematography. ECNDT TH.1.3.2. Fraunhofer Institute for High Speed Dynamics, Ernst-Mach-Institut.

20. Cloens, P. September 18, 2006. *X-Ray Imaging Instrumentation*. ESRF, Grenoble, France.

7

MEMS, MOEMS, Nano, and Bionanotechnologies

7.1 Introduction

The simple distinction between the terms micro-electro-mechanical systems (MEMSs) and nanotechnology is the size of the devices: MEMSs typically range between millimeters down to microns and nano devices are on the nanometer scale. Quoting an excellent set of definitions from SmallTech Consulting [1]:

> MEMS is the integration of a number of microcomponents on a single chip which allows the microsystem to both sense and control the environment. The components typically include microelectronic integrated circuits (the "brains"), sensors (the "senses" and "nervous system"), and actuators (the "hands" and "arms"). The components are typically integrated on a single chip using microfabrication technologies similar to those used for integrated circuits.
>
> Nanotechnology takes advantage of the observation that at the nanoscale, properties of materials change. Nanotechnology is that array of technologies that use properties of materials that are unique to structures at the nanoscale.

A Richard Feynman presentation in 1959 discussed material fabrication "from the atom up," leading to the debate regarding the concept of molecular manufacturing and molecular nanotechnology. The classic Drexler–Smalley debate focused on the feasibility of constructing molecular assemblers [2]. Multidisciplinary efforts continue to pace the complementary fields of MEMS, MOEMS, nano, and bionanotechnology, with exponential growth upon exponential growth, advancing toward Ray Kurzweil's projected "singularity" around the year 2040. Nanomaterials for biology now include quantum rods and dots (CdSe, CdTe, and CdHgTe), non-heavy-metal-based quantum dots (InP, $CuInS_2$, silicon, magnetic nanoclinics, silica nanoparticles, metallic [gold and silver] particles and rods, and rare-earth-doped nanophosphors [Er and $Tm:NaYF_4$]) [3]. A significant segment of the MEMS market is that of MOEMS—optical MEMS—partly a result of the fusion of computing and signal processing with photonics and communications—including the marriage of optics and semiconductor micromachining and devices—playing heavily in the lab on a chip concepts for biomedicine. Certainly, the time is ripe for the convergence of MEMS, nanotechnology, and biotechnology. Perhaps, MOEMS devices interacting with biological systems will manipulate nanoparticles in applications for drug delivery, gene therapy, and commercial labs on a chip.

MOEMS advanced rapidly in part due to the telecommunication market's demand for devices such as add-drop mulitplexers, optical cross connect switches, variable optical attenuators, modulators, dense wavelength division multiplexing, wavelength converters, dynamic gain equalizing filters, tunable filters, and advanced packaging technologies. More than 50 start-up ventures in the late 1990s complemented the efforts of the large telecommunication manufacturers in the acceleration of such market applications.

Significant applications are emerging in the related field of biomimetics, involving biologically inspired synthetic materials. These point to the construction of genetic networks from and within cells, along with intelligent implantable sensors. Microfluidic MEMS constructs will provide processing of antibodies, peptides, metabolic markers, and monitors. DNA may even provide the backbone for computer logic chips. Biosensors mimicking mammalian physiological behavior, fabricated from completely synthetic abiotic materials, can be programmed to sense, respond, and adapt to requirements of living systems, along with providing the basis for a range of chemical and biological sensing systems [4].

With the emergence of multidisciplinary fields such as bioinformatics, mathematical biology, computational biological modeling, biophysics, etc., we should continue to see the cataloging and relational systems approach to biology and the sciences as part of the exponential "Kurzweilian" growth characteristic of technological progress. Crystal growth techniques described in earlier chapters for the III–V GaAs semiconductor material system are analogous in many respects to growth techniques used to nucleate the precipitation of protein crystalline structure from aqueous solution.

Quoting D. A. LaVan and R. Langer, MIT, from the NSF Symposium in 2001:

> While some may dream of nanorobots circulating in the blood, the immediate applications in medicine will occur at the interfaces among... nanotechnology, micro-electronics, microelectromechanical systems (MEMS) and microopticalelectro-mechanical systems (MOEMS)...The bounty will not be realized until those trained in these new paradigms begin to ....address basic medical and scientific questions.

7.2 MEMS and Nanotechnology*

7.2.1 Introduction

The section provides a discussion of MEMS and nanotechnology.

* Courtesy of Dr. Michael Huff of the *MEMS and Nanotechnology Exchange* (see http://mems-exchange.org) at the *Corporation for National Research Initiatives* [5].

MEMS is the integration of mechanical elements, sensors, actuators, and electronics on a common silicon substrate through microfabrication technology. While the electronics are fabricated using IC process sequences (e.g., CMOS, Bipolar, or BICMOS), the micromechanical components are fabricated using compatible "micromachining" processes that selectively etch away parts of the silicon wafer or add new structural layers to form the mechanical and electromechanical devices (see Figure 7.1).

MEMS promises to revolutionize nearly every product category by bringing together silicon-based microelectronics with micromachining technology, making possible the realization of complete systems-on-a-chip. MEMS is an enabling technology allowing the development of smart products, augmenting the computational ability of microelectronics with the perception and control capabilities of microsensors and microactuators, and expanding the space of possible designs and applications.

Microelectronic integrated circuits (ICs) can be thought of as the "brains" of a system, and MEMS augments this decision-making capability with "eyes" and "arms" to allow microsystems to sense and control the environment. Sensors gather information from the environment through measuring mechanical, thermal, biological, chemical, optical, and magnetic phenomena. The electronics then process the information derived from the sensors and through some decision-making capability direct the actuators to respond by moving, positioning, regulating, pumping, and filtering, thereby controlling the environment for some desired outcome or purpose. Because MEMS devices are manufactured using batch fabrication techniques similar to those used for ICs, unprecedented levels of functionality, reliability, and sophistication can be placed on a small silicon chip at a relatively low cost.

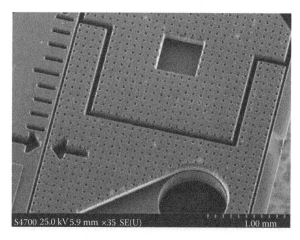

FIGURE 7.1
MEMS structure. (Courtesy of the MEMS and Nanotechnology Exchange.)

There are numerous possible applications for MEMS and nanotechnology. As a breakthrough technology, allowing unparalleled synergy between previously unrelated fields such as biology and microelectronics, many new MEMS and nanotechnology applications will emerge, expanding beyond that which is currently identified or known. Following are a few applications of current interest.

7.2.2 Biotechnology

MEMS and nanotechnology are enabling new discoveries in science and engineering such as the polymerase chain reaction (PCR) microsystems for DNA amplification and identification, micromachined scanning tunneling microscopes, biochips for detection of hazardous chemical and biological agents, and microsystems for high-throughput drug screening and selection.

7.2.3 Communications

High-frequency circuits will benefit considerably from the advent of the RF-MEMS technology. Electrical components such as inductors and tunable capacitors can be improved significantly compared to their integrated counterparts if they are made using MEMS and nanotechnology. With the integration of such components, the performance of communication circuits will improve, while the total circuit area, power consumption, and cost will be reduced. In addition, the mechanical switch, as developed by several research groups, is a key component with huge potential in various microwave circuits. The demonstrated samples of mechanical switches have quality factors much higher than anything previously available.

Reliability and packaging of RF-MEMS components seem to be the two critical issues that need to be solved before they receive wider acceptance by the market.

7.2.4 Accelerometers

MEMS accelerometers are quickly replacing conventional accelerometers for crash air-bag deployment systems in automobiles. The conventional approach uses several bulky accelerometers made of discrete components mounted in the front of the car with separate electronics near the air bag; this approach costs over $50 per automobile. MEMS and nanotechnology have made it possible to integrate the accelerometer and electronics onto a single silicon chip at a cost between $5 and $10. These MEMS accelerometers are much smaller, more functional, lighter, and more reliable and are produced for a fraction of the cost of the conventional macroscale accelerometer elements.

FIGURE 7.2
MEMS structure. (Courtesy of the MEMS and Nanotechnology Exchange.)

MEMS and nano devices are extremely small, for example, MEMS and nanotechnology have made possible electrically driven motors smaller than the diameter of a human hair (see Figure 7.2), but MEMS and nanotechnology are not primarily about size. MEMS and nanotechnology are also not about making things out of silicon, even though silicon possesses excellent materials properties, which make it an attractive choice for many high-performance mechanical applications; for example, the strength-to-weight ratio for silicon is higher than many other engineering materials, which allows very high bandwidth mechanical devices to be realized. Instead, the key importance of MEMS and nano is as a new manufacturing technology—a way of making complex electromechanical systems using batch fabrication techniques similar to those used for ICs and uniting these electromechanical elements together with electronics.

7.2.5 Advantages of MEMS and Nano Manufacturing

First, MEMS and nanotechnology are extremely diverse technologies that could significantly affect every category of commercial and military product. MEMS and nanotechnology are already used for tasks ranging from in-dwelling blood pressure monitoring to active suspension systems for automobiles. The nature of MEMS and nanotechnology and its diversity of useful applications make it potentially a far more pervasive technology than even IC microchips.

Second, MEMS and nanotechnology blur the distinction between complex mechanical systems and IC electronics. Historically, sensors and actuators are the most costly and unreliable part of a macroscale sensor-actuator-electronics system. MEMS and nanotechnology allow these complex electromechanical systems to be manufactured using batch fabrication techniques, decreasing the cost, and increasing the reliability of the sensors and actuators to equal those of ICs. Yet, even though the performance of MEMS and nano devices is expected to be superior to macroscale components and systems, the price is predicted to be much lower.

MEMS technology is based on a number of tools and methodologies, which are used to form small structures with dimensions in the micrometer scale (one millionth of a meter). Significant parts of the technology have been adopted from IC technology. For instance, almost all devices are built on wafers of silicon, like ICs. The structures are realized in thin films of materials, like ICs. They are patterned using photolithographic methods, like ICs. There are however several processes that are not derived from IC technology, and as the technology continues to grow the gap with IC technology also grows.

There are three basic building blocks in MEMS technology, which are the ability to deposit thin films of material on a substrate, to apply a patterned mask on top of the films by photolithographic imaging, and to etch the films selectively to the mask. A MEMS process is usually a structured sequence of these operations to form actual devices.

7.2.6 Developments Needed

MEMS and nanotechnology are currently used in low- or medium-volume applications.

Some of the obstacles preventing its wider adoption are the following.

7.2.6.1 Limited Options

Most companies who explore the potential of MEMS and nanotechnology have very limited options for prototyping or manufacturing devices and have no capability or expertise in microfabrication technology. Few companies will build their own fabrication facilities because of the high cost. A mechanism giving smaller organizations responsive and affordable access to MEMS and Nano fabrication is essential.

7.2.6.2 Packaging

The packaging of MEMS devices and systems needs to improve considerably from its current primitive state. MEMS packaging is more challenging than IC packaging due to the diversity of MEMS devices and the requirement that many of these devices be in contact with their environment. Currently, almost all MEMS and nano development efforts must develop a new and specialized package for each new device. Most companies find that packaging is the single most expensive and time-consuming task in their overall product development program. As for the components themselves, numerical modeling and simulation tools for MEMS packaging are virtually nonexistent.

Approaches that allow designers to select from a catalog of existing standardized packages for a new MEMS device without compromising performance would be beneficial.

7.2.6.3 Fabrication Knowledge Required

Currently, the designer of a MEMS device requires a high level of fabrication knowledge in order to create a successful design. Often, the development of even the most mundane MEMS device requires a dedicated research effort to find a suitable process sequence for fabrication. MEMS device design needs to be separated from the complexities of the process sequence.

7.3 Nanotechnology Applications

Figure 7.3 from the Foresight Nanotech Institute [6] represents the technology roadmap for the key application areas of nanotechnology. This chart illustrates the technology convergence and rapid progression to a host of diverse applications and implementations. As the theme of this text and the cover art suggest, progress in optical bionanotechnology will follow cross-disciplinary endeavors in achieving some of the most significant breakthroughs in science.

Applications in biotechnology include bioimaging, biosensors, flow cytometry, photodynamic therapy, tissue engineering, and bionanophotonics. Breakthroughs in biological signaling, genomics, and biosystems engineering will follow as well.

7.4 V-Groove Coupler Geometry and Design Considerations

Bulk micromachining of silicon and other semiconductor materials represents the early work in the MEMS arena. A very early application of anisotropically etched silicon was the V-groove coupler. Figure 7.4 illustrates a round fiber in a V-groove, providing a very precise means of aligning a fiber to another fiber or for coupling to waveguides, detector elements, or other optical components. We consider here the geometry and design considerations for such a structure.

It is found that the problem of finding b, the location of the center of the fiber with respect to the V-groove, is dependent upon whether the point of contact is at the top corner of the groove or on the side wall. The point of contact is described by Ψ, c', and d'. Then, $c'/r = \cos \Psi$.

The value of b sat this point is defined as

$$b\left(\frac{c'}{r} = \cos \Psi\right) = b_0 = r \sin \Psi \qquad (7.1)$$

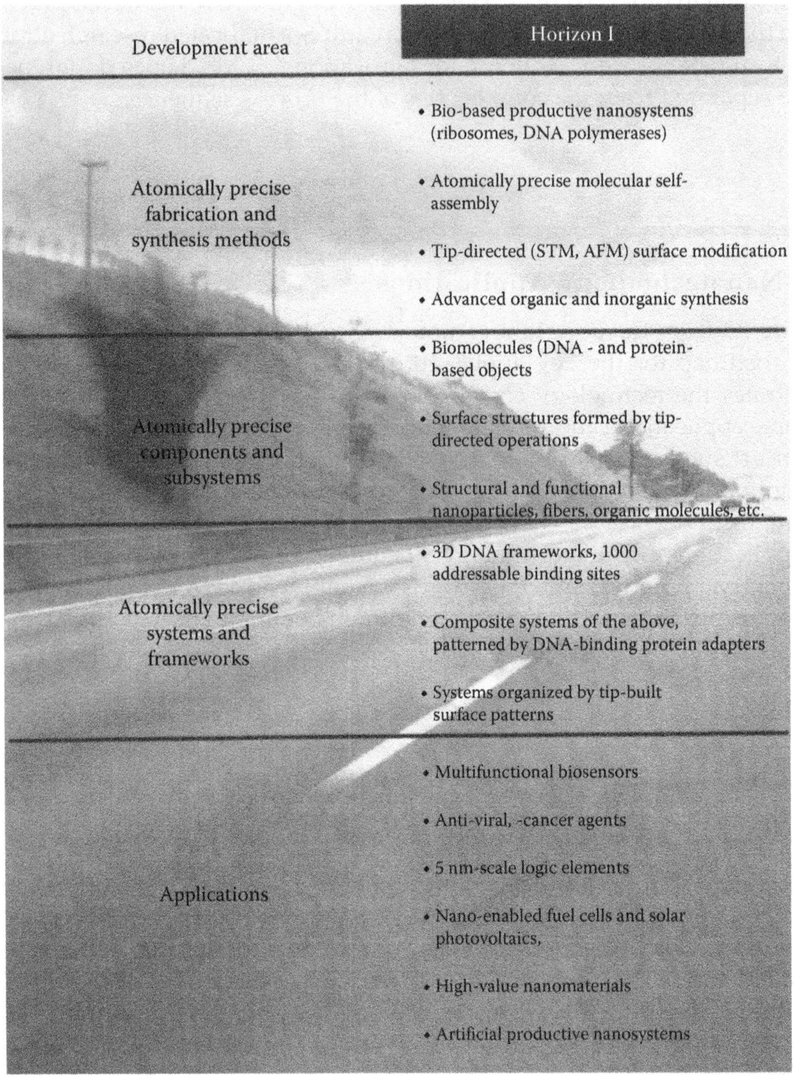

Development area	Horizon I
Atomically precise fabrication and synthesis methods	• Bio-based productive nanosystems (ribosomes, DNA polymerases) • Atomically precise molecular self-assembly • Tip-directed (STM, AFM) surface modification • Advanced organic and inorganic synthesis
Atomically precise components and subsystems	• Biomolecules (DNA - and protein-based objects • Surface structures formed by tip-directed operations • Structural and functional nanoparticles, fibers, organic molecules, etc.
Atomically precise systems and frameworks	• 3D DNA frameworks, 1000 addressable binding sites • Composite systems of the above, patterned by DNA-binding protein adapters • Systems organized by tip-built surface patterns
Applications	• Multifunctional biosensors • Anti-viral, -cancer agents • 5 nm-scale logic elements • Nano-enabled fuel cells and solar photovoltaics, • High-value nanomaterials • Artificial productive nanosystems

(a)

FIGURE 7.3
Nanotechnology roadmap. (Courtesy of the Foresight Nanotech Institute, Palo Alto, CA.)

Horizon II	Horizon III
• Artificial productive nanosystems in solvents	• Scalable productive subsystems in machine-phase environments
• Mechanically directed solution-phase synthesis	• Machine-phase synthesis of exotic structures
• Directed and conventional self-assembly	• Multi-scale assembly
• Crystal growth on tip-built surface patterns	• Single-product, high-throughput molecular assembly lines
• Coupled-catalyst systems	
• Composite structures of ceramics, metals, and semiconductors	• Nearly reversible spintronic logic
• Tailored graphene, nanotube structures	• Microscale 1 MW/cm³ engines and motors
• Intricate, 10 nm scale functional devices	• Complex electro-mechanical subsystems
	• Adaptive supermaterials
• Casing, "circuit boards" to support, link components	• Complex systems of advanced components, micron to meter + scale
• 100 nm scale, 1000-component systems	• 100 GHz, 1 Gbyte, 1-Ψm-scale, sub-ΨW processors
• Molecular motors, actuators, controllers	• Ultra-light, super-strength, fracture-tough structures
• Digital logic systems	
• Artificial immune systems	• Artificial organ systems
• Post-silicon extension of Moore's law growth	• Exaflop laptop computers
• Petabit RAM	• Efficient, integrated, solar-based fuel production
• Quantum-wire solar photovoltaics	• Removal of greenhouse gases from atmosphere
• Next-generation productive nanosystems	• Manufacturing based on productive nanosystems

(b)

FIGURE 7.3 (continued)

The problem is now dependent upon the value of c/r. For values of c/r such that $c/r \leq \cos \Psi$, we term the value of b as $b_<$:

$$b_<^2 = (r^2 - c^2) \quad \text{or} \quad b_< = r \left[1 - \left(\frac{c}{r} \right)^2 \right]^{\frac{1}{2}} \tag{7.2}$$

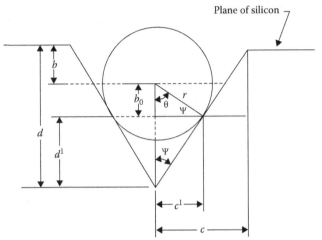

r = Radius of cylinder (fiber)
c = 1/2 Width of the groove
d = Depth of the groove
Ψ = 1/2 Angle of the "V"
θ = 1/2 Angle of the points of
 contact cylinder and groove
b = Height of cylinder center above plane
c = Half width of the groove

FIGURE 7.4
V-groove geometry.

Assuming Ψ constant, we look for variations in $b_<$ caused by variations in c.

$$\frac{db_<}{dc} = \frac{1}{\left[(r/c)^2 - 1\right]^{\frac{1}{2}}} \tag{7.3}$$

For values of c/r, such that $c/r > \cos \psi$, we term the value of b as $b_>$:

$$b_> = b_0 - (d - d')$$

The values of d and d' are

$$d = \frac{c}{\tan \psi} \quad \text{and} \quad d' = \frac{c_0^2}{b_0} = \frac{r^2}{b_0} - b_0 \tag{7.4}$$

Substitution yields

$$b_> = \frac{r}{\sin \psi} - \frac{c}{\tan \psi} \quad \text{or} \quad b_> = \frac{r}{\sin \psi}\left(1 - \frac{c}{r}\cos \psi\right) \tag{7.5}$$

It is found that the variation of $b_>$ with changes in c is constant:

$$\frac{db_>}{dc} = \frac{-1}{\tan \psi} \tag{7.6}$$

Figure 7.5 illustrates the form of Equations 7.3 and 7.6 for an arbitrary value of ψ. This analysis shows that the vertical alignment improves for decreasing values of c/r. On the other hand, it is found that the percent of exposed fiber, expressed for convenience as the percent of the fiber diameter above the surface of the chuck, decreases for increasing values of c/r. This percentage is a parameter that can be indicative of the quality of the horizontal stability of the fiber alignment, with lower values indicating better stability. It is seen that better vertical alignment is obtained to some degree at the expense of this horizontal stability.

As a numerical example, we use the value of $\psi = 36°$. This is the value for the anisotropic etching angle in silicon for coupling chucks. The table at Figure 7.6 provides results for several values of c/r. Point 1 shows that perfect vertical alignment is only obtained with an infinite horizontal uncertainty as there is of course no groove ($c = 0 = c/r$). At point 2, there is still a great deal of the fiber above the plane of the chuck. Point 3 shows that, even when there can be no further increase in the quality of vertical alignment, there is still 80% exposure. It therefore appears necessary to sacrifice some vertical accuracy in order to obtain a reasonable amount of horizontal stability. It appears that the most reasonable value of c/r to use is in the vicinity of point 4 where the fiber core is aligned near the surface of the chuck.

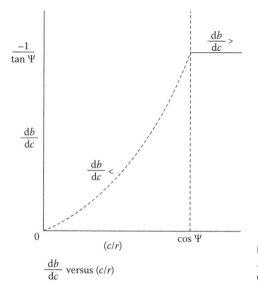

$\frac{db}{dc}$ versus (c/r)

FIGURE 7.5
Alignment error as a function of groove dimension.

Point	$\dfrac{c/r}{\cos \Psi}$	$\dfrac{db}{dc}$	$\dfrac{b}{r}$	% Fiber exposed
1	0	0	1.00	100
2	0.50	−0.44	0.91	95
3	1.00	−1.38	0.59	80
4	1.53	−1.38	0	50
5	2.00	−1.38	−0.53	24

FIGURE 7.6
Vertical misalignment (db/dc), height of fiber core above chuck plane (b/r), and percent of exposed fiber as functions of groove width to fiber diameter ratio (c/r).

To obtain an idea of the actual sizes of the misalignment, we consider a total groove width error of +0.5 μm. Then, the center of the fiber will be lower than expected by −0.69 μm. This would be an important consideration for direct integration of grooves and waveguides, considering a single-mode core diameter of perhaps 4 μm; however, many efforts are directed toward separately aligning the chuck and guides, so the real interest is in keeping a consistent groove width among multiple grooves. One might expect a plus or minus 0.1 μm variance between three grooves, and, therefore, the three fiber cores would be aligned to ±0.14 μm.

Consider two fiber diameters of 80 and 125 μm. By aligning the core of the smaller fiber with the chuck surface,

$$b_> = 0 \text{ and therefore } \frac{c}{r} = \frac{1}{\cos \psi}$$

Using the values $r = 40\,\mu m$ and $\psi = 36°$,

$$c = \frac{40\ \mu m}{\cos 36} = 49.4\ \mu m$$

The critical values for both fibers are summarized in Figure 7.7.

Some representative V-groove chuck dimensions are shown at Figures 7.8 and 7.9. The groove width would be the critical dimension and would determine the groove depth by the crystal geometry. The groove width may be 96–100 μm as long as each of the three grooves is the same. The 200.6 μm groove separation is also very important. This basic approach represents the basis for a range of early silicon-bench optical-packaging devices.

Fiber-Dimensions		V-Groove Values				
Diameter	Radius (r)	c (μm)	d (μm)	c/r	c (μm)	b/r
80 μm	40 μm	49	67.4	1.225	0.6	0.015
125 μm	62–5 μm	49	67.4	0.789	38.9	0.620

FIGURE 7.7
Critical values for 80 and 125 μm diameter fibers.

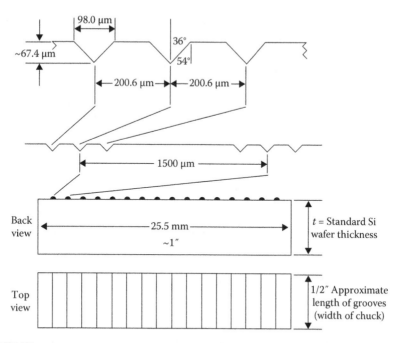

Back view

Top view

FIGURE 7.8
V-groove chuck for 80 and 125 µm diameter fibers.

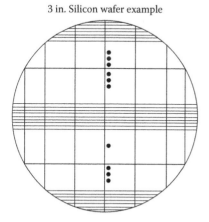

3 in. Silicon wafer example

FIGURE 7.9
Cleavage lines yielding about 14 useful chucks, set of 3 grooves, set spacing, 1.5 mm indicate continued grooving.

7.5 Bionanotechnology

7.5.1 Introduction

Nanoscale scaffolds may ultimately be utilized for molecular structuring (ala the classic Drexler versus Smalley debate regarding molecular building blocks and how they might be manipulated) as well as for regulating information transfer. Ron Weiss's page at Princeton [7] describes building biologic circuits. SPICE electrical circuit simulations are now complemented with bioSPICE for modeling genetic circuits [8]. Biological system simulation programs are available as open source code.

Molecular recognition is a key objective for a host of multiplexed biosensors technologies in development. These include biomedical MEMS cantilever sensors, surface effects on nanocantilevers, and cancer cantilever sensors. Surface acoustic wave (SAW) sensors prove quite sensitive to adsorbed materials on the device surface, and a wide range of electrochemical sensors detects free radicals, glucose, and nitric oxide in tissues.

Photonic nanosensors monitor biochemical reactions through conversion of chemical energy to light signals. This is now accomplished for antibody detection inside a single cell [9]. Charge density waves resulting from light impinging on the interface between a thin film and a second medium, called surface plasmons, are utilized to study the formation of self-assembled layers and molecular structures, including DNA and proteins.

7.5.2 Biology Labs on a Chip

Integrated biochips are now commercially available, integrating biotechnology with MEMS, optics and electronics, imaging, and processing in a miniaturized hybrid system [10]. The term biochip is synonymous in this context with the IUPAC definition of biosensor: "a self-contained integrated device, which is capable of providing specific quantitative or semiquantitative analytical information using a biological recognition element (biochemical receptor), which is retained in direct spatial contact with a transduction element."

The general concept for analysis of a sample (analyte) is for the bioreceptor to provide molecular recognition of the DNA, RNA, or protein molecule and to affect a signal through a transducer that provides a measurement or characterization output. Biochips are distinguished from microarray assay systems, in that the biochip contains both the microarray and the detector array transduction system. While array plates provide the ability to identify analytes with very high volume, speed, and throughput, biochips afford portability, lower cost for small batch processing, and automated analysis algorithms. This is more convenient for applications such as field monitoring or physician's office testing.

Multifunction biochips provide the ability to detect DNA, RNA, protein, and specific characteristics unique to the biological probe composition, such as specific enzymes, antibodies, antigens, and cellular structures [11]. Transducers vary widely, such as surface plasmon resonance devices, optical property measurement devices (absorption, luminescence), electrochemical, SAW, and other mass measurement devices. Progress in medicine will rely heavily upon the biochip systems currently in development [12].

7.5.3 Applications in Bioorganic Chemistry

7.5.3.1 Smell

Carvone exists as a pair of enantiomers. (R)-(−)-carvone smells like spearment, whereas (S)-(+)-carvone smells like caraway. Why do these enantiomers have different smells (i.e., different biological activity)? Olfactory receptors in the cell membranes of humans apparently possess the ability to respond differently to the two carvone enantiomers. Considering each cell as an analog computer, processing fuzzy logic [13], the reaction of the receptor proteins in the olfactory complex has the ability to regulate ion transport through the cell membrane differentially for the two carvone enantiomers, thereby producing a different electrochemical signal to the brain. Perhaps, the selection processes in mitochondrial evolution resulted in certain specific cellular logic functions for smell, and specific abilities of species to distinguish and differentiate enantiomers such as carvone [14]. The two enantiomers "present themselves" differently to the olfactory receptors, resulting in a differential response. It is also possible that the two enantiomers present a different fit to the hydrogen-bonded water networks that play such an important role in "presenting" molecules to one another in biological solution. Complementary shapes are thus selected for a specific reaction.

7.5.3.2 Protein Structure and Folding and the Influence of the Aqueous Environment

Analysis and characterization of proteins represent an important branch of biotechnology. The theories of hydrogen bonding are proving of great importance to the understanding of protein behavior. Martin Chapman's site [15] describes extended hydrogen-bonded water molecules, such as icosahedrals, which should be visible under a 1000× or 2000× microscope, if indeed they are stable and exist for more than the expected femtosecond duration.

Further experimentation is needed to properly characterize these structures. It may be possible for water to remain in an aggregate represented by one of Chaplin's "magic numbers" for more than a very short duration. This is a bit controversial in the literature. Some of the abstracts from a recent conference on water [16] indicate that indeed this occurs. If so, perhaps we can control the behavior of protein folding in "conditioned" structured aqueous environments. Another site also discusses water as a "designer fluid" that helps proteins change shape [17].

Protein folding helps determine the functionality of the biomacromolecule. Aspartic acid, for example, is considered acidic hydrophilic [18]. When the protein folds, the side chains can be transferred from a solution-like environment to one determined by the 3-D structure of the folded protein. So, exposure to solvents could be impacted differentially on one side of the protein versus the other. Differential interaction of the aspartic acid side chain with permanent charges and protein dipoles will also affect the pH of the side chain [19]. Exposure to the physiological conditions of the solvent is affected by the amount of shielding by the protein fold—different on one side versus the other. Specific biological function could be impacted and determined by the solvent-determined activity of the side chains. Those side-chain pH's could be determined by their ability to interact with the solvent.

7.5.3.3 Evolution and the Biochemistry of Life

When formed synthetically, amino acids contain equal amounts of L and D enantiomers:

"Amino acids synthesized in the laboratory are a mixture of the right-and left-handed forms, and thermodynamically the two forms are indistinguishable" [20]. This fact rather disproved the famous Miller experiments of 1953, in which traces of nonliving amino acids were created in a mix of laboratory gases and even represented one of the holes in the evolutionary origin-of-life theories [21].

One theory is that electroweak interactions of electromagnetism and the weak nuclear force, which calculations suggest, will result in a difference in the L and D enantiomers of one part in 10^{-17}, which eventually dominated, although this theory is about as convincing (not very) as the idea that earth's first biopolymers were delivered to earth by meteorites. The lack of convincing experimental (or fossil) support leaves these as unproven theories [22].

If the amino acids in meteorites are used as the basis to represent those available on the primitive earth, we must account for methylation into a form that is not biologically useful. Ron Breslow and Mindy Levine at Columbia University determined that the very small imbalance in favor of the L enantiomer can be amplified to 90% L through just several cycles of evaporation. Also, the presence of copper in the primordial stew results in a strong initial L bias [23].

Equally intriguing and unproven theories include the delivery of "structured water" via meteorites to the ancient Dogon Nomos in Africa. Legend has it this water delivered the "ancient message" and contained extraterrestrial communications that provided man (through biochemistry) with intelligence. This may be the intelligence in Genesis that separated the robotic, programmed humanoids from the reasoning man with a soul, ala "The Origins of Consciousness in the Breakdown of the Bi-cameral Mind" [24].

It would seem that there is only one unique solution to the biochemistry of life. The 20 amino acids must occur in the correct sequence, in the correct form, for life to exist in the protein structures. Because of the incredibly complex series of interactions needed for the process of life, the wrong orientation would upset the entire logical mesh represented in the human form. While arguments for primordial bias due to volcanic conditions, heat for evaporation, and isolated copper-laden lakes might provide the basis for the L configuration bias and therefore an explanation for the specific L role in biological complexity, it seems that there was ample opportunity for the complementary enantiomer to evolve into life forms as well—if indeed, it provided an eigenvalue for mathematical resolution.

A single error in the amino acid sequence can manifest itself in genetic dysfunction and death. Astounding as it is to comprehend the complexity of life, it would seem that we are dealing with the simultaneous solution to a host of equations, with a single unique solution. It is possible that the first self-replicating protein structure could have started with the D structure, but it apparently did not. Panspermia, regional specificity, meteorites, etc. may, in totality, represent the initial bias, but systems biology and mathematical uniqueness may be the ultimate determinant. Perhaps, even the influence of aggregated water on the deep major and minor grooves has a dominant unique solution. Absence of proof implied proof of lack of a physically realizable solution to an alternate life form. The 1% bias theories are interesting, but probably not necessary for the real solution we have realized in our biochemistry.

7.5.3.4 Biochemical Analysis and Cancer

Jonsson et al. published "A strategy for identifying differences in large series of metabolomic samples analyzed by GC/MS" [25] in which they describe the process for identification of metabolites in a biological system, using standard techniques of organic chemistry including gas chromatography, mass spectrometry, ^{1}H NMR, and data-processing algorithms for rapid analysis of large data sets. Additional organic chemistry analytical methods applied to metabolomics include chromatography, electrophoresis, and mass spectroscopy. Thus, the traditional techniques of organic chemistry are applied to biological systems of metabolites, hormones, and signaling mechanisms in the emergent field of metabolomics.

Metabolmics provides a complete understanding of cell physiology, complemented by the studies of related genomics and proteomics. This approach was first applied to improved understanding of metabolism [26]. It is thought that a systematic cataloging of the human metabolome will provide a baseline for studies of cellular processes and their perturbations.

The Human Metabolome Database database, supported by the University of Alberta [27] and Genome Canada, contains links to chemical, clinical, and biochemical/molecular biological data, with links to protein and DNA sequences. This systems approach to biological studies promises rapid streamlining of previously incredibly tedious processes.

For instance, how much additional progress would have resulted in the studies of Dr. Judah Folkman on angiogeniesis and its relationship to cancer, had the tools of metabonomics been available a few years ago [28]? Evidence of the growing success of the field is demonstrated in part by growing conferences and attention in the literature to metabolomics, following the first mention of the approach about 10 years ago.

Soon perhaps, a treatment for cancer epigenetics will be created through localized solvent physiology changes that reprogram the cancer cell behaviors. If the triggering mechanism for cancer cell programming is identified and reversed, the basis for a cure or reversal of the condition may be possible. Antiangiogensis agents, epigenetic reprogramming, perhaps even structured water with appropriate solvents changing the ion transport of potassium across cell boundaries, may alter sufficiently the folding of proteins during the cancer cell processes. It seems that recent breakthroughs [29] in the understanding of DNA methylation may provide the epigenetic key for screening, prevention, and cure of many or all cancers.

7.5.4 Bioimaging Applications

Optical imaging techniques include fluorescence microscopy, Raman imaging, interference imaging, optical coherence tomography, total internal reflection imaging, multiphoton microscopy, confocal microscopy, and other developing tools including fluorescence imaging. Fluorescence optical imaging systems include spatial filtering confocal microscopy, spatially resolved localized spectroscopy, polarization and time resolved fluorescence lifetime imaging, and fluorescence resonance energy transfer. Applications include whole body imaging, drug distribution, protein engineering, and identification of structural changes in cells, organelles, and tissues [30].

X-ray crystallography provides another means to characterize the atomic structure of crystalline materials. (Rosalind Franklin used the technique to produce the famous demonstration that DNA is helical.) This characterization includes proteins and nucleic acids. While certain proteins are difficult to crystallize, nearly 50,000 proteins, nucleic acids, and other biological macromolecules have now been measured with x-ray crystallography [31,32].

7.5.5 Biologically Inspired Computing and Signal Processing

Interesting switching behaviors occur in proteins [33]. There are aspects of epigenetic programming that may possibly be "switched" with light. If we express such a protein in a cell, we could control aspects of the protein's behavior with light. We must determine what controls and expresses the switching information and what activates it. Some work has been accomplished with cutting and pasting proteins to create light-sensitive kinases. Adding and subtracting amino acid residues modulate the activity [34,35].

As discussed in Chapter 1, this is analogous to the progression of (electrical) signal processing using combinations of digital logic gates and application-specific ICs; to optical signal processing using operator transforms, optical index, and nonlinear bistable functions through acousto- and electro-optic effects to perform optical computing; to Dennis Bray's "Wetware" biologic cell logic; and to the biophilosophical, astounding insights of Nick Lane in the last two chapters of his recent "Life Ascending- the Ten Great Inventions of Evolution" [36], where Lane characterizes the evolutionary development of consciousness and death (through the mitochondria), providing the basis for biological logic and how best to modify and control that inherent logic.

We are now on the cusp of an epoch, where it is possible to design light-activated genetic functionality, to actually regulate and program biological functionality. What if, for instance, we identify/isolate the biological "control mechanism" for the switching on/off of limb regeneration in the salamander, or, with the biophotonic emission associated with cancer cell replication and analogous to the destruction of the cavity resonance in a laser, we reverse the biological resonance and switch or epigenetically reprogram the cellular logic? Soon we will engineer time-sequenced reaction pathways to regulate and affect a huge variety of complex biochemical functionalities [37,38].

The emerging field of optogenetics addresses how specific neuronal cell types contribute to the function of neural circuits, the idea that two-way optogenetic traffic could lead to "human–machine fusions in which the brain truly interacts with the machine rather than only giving or only accepting orders..." [39]. This could be Ray Kurzweil's "singularity"; he said that it is "near" (about the year 2040) in his documentary "Transcendent Man." Device physics and electro-optic/electrical engineering, optical MEMS ("MOEMS the word!"), and bionanotech have converged!

Protein scaffolds bind core kinases that successively activate one another in metabolic pathways engineered to provide biological signal processing functions. Envision three-dimensional nanostructures fabricated using laser lithography. Then assemble complex protein structures onto these lattices in a preferred manner, such that introduction of the lattice frames might "seed" the correct protein configuration. We then conceive biocompatible nanostructures for "biologically inspired" computing and signal processing, for biosensors in the military, for example, or for recreating two-way neural processes and repairs.

By spatially recruiting metabolic enzymes, protein scaffolds help regulate synthetic metabolic pathways by balancing proton/electron flux. It seems that this represents a synthesis methodology with advantages over more standard chemical constructions. We used to design sequences of digital optical logic gates to remove the processing bottlenecks of conventional computers. But if we look to producing non-Von Neumann processors with biological constructs, we may truly be at the computing crossroad for the "Kurzweil singularity"!

References

1. Small Tech Consulting. MEMS and nanotechnology. http://www.smalltechcon sulting.com/What_are_MEMS_Nanotech.shtml

2. Bueno, O. 2004. The Drexler-Smalley debated on nanotechnology: Incommensurability at work? *Hyle-Int. J. Philos. Chem.* 10(2):83–98.
3. Prasad, P. N. 2003. *Introduction to Biophotonics.* Hoboken, NJ: John Wiley & Sons.
4. Valdes, J., E. Valdes, and D. Hoffman. 2009. Towards complex abiotic systems for chemical and biological sensing. Final report to Aberdeen Proving Ground, ECBC-TR-720. Approved for public release and unlimited distribution.
5. MEMS and Nanotechnology Exchange. http://www.mems-exchange.org/
6. Foresight Nanatech Institute. 2007. Productive nanasystems, a technology roadmap. UT-Battelle, LLC, Publisher.
7. Weiss, R. 2005. Princeton University Department of Electrical Engineering. http://www.ee.princeton.edu/people/Weiss.php
8. Bio-SPICE. Biological simulation program for intra- and inter-cellular evaluation. http://biospice.sourceforge.net/
9. Hornyak, G., J. Moore, H. Tibbals, and J. Dutta. 2009. *Fundamentals of Nanotechnology.* New York: Taylor & Francis Group.
10. Vo-Dinh, T. 2003. *Biomedical Photonics Handbook.* Boca Raton, FL: CRC Press.
11. Vo-Dinh, T. and M. Askari. 2001. Micro-arrays and biochips: Applications and potential in genomics and proteomics. *Curr. Genomics* 2:399.
12. Prasad, P.N. *Introduction to Biophotonics.* New York: Wiley.
13. Bray, D. 2009. *Wetware: A Computer in Every Living Cell.* New Haven: Yale University Press.
14. Lane, N. 2006. *Power, Sex, Suicide: Mitochondria and the Meaning of Life.* Oxford: Oxford University Press.
15. Chaplin, M. Water structure and science http://www1.1sbu.ac.uk/water/models.html. Accessed December 2010.
16. Schedule and speakers. *Conference on Physics, Chemistry & Biology of Water.* http://www.watercon.org/schedule.html
17. University of Illinois at Urbana-Champaign. 2008. Water is "designer fluid" that helps proteins change shape. Science Daily. http://www.sciencedaily.com/releases/2008/08/080806113314.htm
18. File: L-aspartic acid skeletal.png. http://en.wikipedia.org/wiki/File:L-aspartic-acid-skeletal.png
19. Protein pKa calculations. http://en.wikipedia.org/wiki/Protein_pKa_calculations.
20. Blum, H. 1970. *Time's Arrow and Evolution.* Princeton, NJ: Princeton University Press.
21. Dembski, W. and J. Wells. 2007. *The Design of Life: Discovering Signs of Intelligence in Biological Systems.* Dallas, TX: Foundation for Thought and Ethics.
22. Haynie, D. 2001. *Biological Thermodynamics.* New York: Cambridge University Press.
23. 2008. The origin of life: Not that sinister. *The Economist* April 12.
24. Jaynes, J. 1976. *The Origin of Consciousness in the Breakdown of the Bicameral Mind.* Boston, MA: Houghton Mifflin.
25. National Center for Biotechnical Information. A strategy for identifying differences in large series of metabolomic samples analyzed by GC/MS. U.S. National Library of Medicine. http://www.ncbi.nlm.nih.gov/pubmed/15018577
26. Nicholson, J. K., J. C. Lindon, and E. Holmes. 1999. Metabonomics: Understanding the metabolic responses of living systems to pathophysiological stimuli via multivariate statistical analysis of biological NMR spectroscopic data. *Xenobiotica* 11:1181–1189.

27. Human metabolome database version 2.5. Genome Alberta. http://www.hmdb. ca/
28. Cooke, R. 2001. *Dr. Folkman's War*. New York: Random House.
29. Medical College of Georgia. 2009. Cancer's distinctive pattern of gene expression could aid early screening and prevention. *Science Daily*. http://www. sciencedaily.com/releases/2009/07/090727110641.htm
30. Prasad. *Introduction to Biophotonics*.
31. Scapin, G. 2006. Structural biology and drug discovery. *Curr. Pharm. Des.* 12(17):2087. doi:10.2174/138161206777585201.PMID 16796557.
32. Lundstrom, K. 2006. Structural genomics for membrane proteins. *Cell. Mol. Life. Sci.* 63(22):2597. doi:10.1007/s00018-006-6252-y. PMID 17013556
33. Photonics Media. Light reveals neuron function. http://www.photonics.com/ Content/ReadArticle.aspx?ArticleID=39843
34. Thompson, E. Professor at Johns Hopkins University. Spring 2010. Private conversation with author.
35. Moglich, A., R. A. Ayers, and K. Moffat. February 2009. Design and signaling mechanism of light-related histidine kinases. *J. Mol. Biol.* 385(5):1433–1444.
36. Lane. *Power, Sex, Suicide*.
37. Bashor, C. J., N. C.Helman, S. Yan, and W. A. Lim. 2008. Using engineered scaffold interactions to reshape MAP kinase pathway signaling dynamics. *Science* 319(5869):1539–1543.
38. Dueber J. E., G. C. Wu, G. R. Malmirchegini, T. S. Moon, C. J. Petzold, A. V. Ullal, K. L. Prather, and J. D. Keasling. 2009. Synthetic protein scaffolds provide modular control over metabolic flux. *Nat. Biotechnol.* 27(8):753–759.
39. Chorost, M. November 2009. Powered by photons. *Wired*.

Index

Printed and bound by CPI Group (UK) Ltd, Croydon, CR0 4YY

21/10/2024

01777107-0011